N & N Science Series

Earth Science

REVISED 1995 EDITION

Authors:

Wayne H. Garnsey

&

Virginia Page
John Jay High School
Hopewell Junction, New York

Illustrations and Graphics:
Eugene B. Fairbanks
Wayne Garnsey

Cover Design:
Eugene B. Fairbanks

N & N Publishing Company, Inc.
18 Montgomery Street   Middletown, New York 10940
(914) 342 - 1677

Dedicated

to our students, with the sincere hope that our book
will further enhance their education and better prepare them
with an appreciation and understanding
of the scientific principles that shape our world.

Special Credits

*Thanks to the many teachers
who contributed their knowledge, skills, and
years of experience to the making of our review text.*

To these educators, our sincere thanks
for their assistance in the preparation of this manuscript:

**Terence Cunningham
Cindy Fairbanks
F. Tracy Fitchett
James McAden
Anne McCabe
Gloria Tonkinson**

Special thanks to our understanding families.

N & N Science Series — Earth Science has been produced on a Macintosh computer. The Apple applications MacWrite II and MacDraw II were used to produce text and graphics. Original line drawings were reproduced on a Microtek MSF-300ZS scanner and modified with Photoshop. The format, special designs, graphic incorporation, and page layout were accomplished with Ready Set Go! by Manhattan Graphics. Special technical assistance was provided by Frank Valenza and Len Genesee of Computer Productions, Newburgh, New York. To all, thank you for your excellent software, enduring hardware, and technical support.

© Copyright 1985, 1995
N & N Publishing Company, Inc.

SAN # - 216-4221   ISBN # - 0935487-10-7

6 7 8 9 0 BookMart Press 0 9 8 7 6 5 4

Table Of Contents

Vocabulary To Be Understood In Topic I

Classification	Measurements
Density (mass/volume)	Observation
Error	Percent Error
Gases	Phases of Matter
Inferences (interpretations)	Senses
Instruments	Solids
Liquids	Time
Mass	Volume

A. Local Environment

How can the local environment be observed?

Observations involve the interaction of a person's senses, such as sight, hearing, taste, touch and smell, with the environment. Observations directly involve **sensory perception**. Since our powers of observation are limited by our senses, it is often necessary for an observer to use instruments to extend his or her ability to observe and collect data. Any piece of information determined directly through the senses is called an **observation**.

Instruments may be used to improve upon one's powers of observation. For example, it is possible to observe stars with the human eye, but a telescope increases the amount of detail that can be observed.

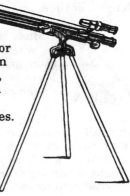

Instruments may be used to extend one's ability to collect data, which cannot be detected by human senses. For example, the human body has no sense receptors which can determine the presence of potentially dangerous radiation, such as x-rays and atomic radiation. However, the use of a detection device like the Geiger counter allows a person to make observations, beyond the capabilities of human senses.

Inferences are interpretations based on observable properties. In other words, an inference is an **interpretation** or a **conclusion** (hypothesis or "educated guess") based on data collected by observation and other pre-gathered information. For example, students observe a smooth stone along a stream. They infer that the smoothness is due to the eroding action of water on the stone. This inference *may* or *may not* be a fact.

In the study of Earth Science, inferences may become "facts" due to the discovery of additional collaborating evidence. The same scientist, or another scientist, may make observations which prove that the water did cause the stone to become smooth.

It is important to understand the difference between an observation (what is discovered with the senses) and an inference (what is interpreted by the mind). Without sufficient observations, inferences may lead to incorrect conclusions.

How can observations of the environment be classified?

A system of classification is based on the properties of the observed event or object. Events or objects that are similar in their properties are generally grouped (organized) together, allowing for more meaningful study. For example, all naturally formed solid objects composed of one or more minerals may be classified as rocks. This helps to organize the study of earth materials. However, it does not accurately describe all rock types. Therefore, rocks are "sub-classified" into many other groups, such as sedimentary, igneous, and metamorphic rocks. In turn, these rock types are again "sub-classified" into further, more distinct groups. For example, sedimentary rocks may be classified as clastic, evaporites, or organic rocks.

B. Environmental Properties

How can properties of the environment be measured?

Units Of Measurements

Measurements contain at least one basic dimensional quantity and describe the properties of objects numerically. These include quantities such as mass, length, or time.

Mass is the amount (quantity) of matter which an object contains and is usually measured in **grams** ($\frac{1}{1000}$ kilogram). Mass is distinguished from weight and should not be confused with it.

Weight of an object is determined by the pull of gravity on the object. Therefore, the object's (e.g., dog's) mass remains constant regardless of the amount of gravitation acting upon it; whereas, the weight of the object (e.g., dog) varies according to the gravitational force.

Length is the distance between the ends or sections of an object, or the total distance between two determined points of that object, and is usually measured in **meters, centimeters** ($1/100$ meter), or **millimeters** ($1/1000$ meter).

Time may be a less definite measurement since it is often a relative event. It is the measurable period during which an action, process, or condition exists, continues, or occurred. It may be a measurement such as a second, or a season, a schedule, an age, or a generation.

Dimensional Quantities And Comparisons

Some properties of matter cannot be measured by basic, single units of measurement. Instead, mathematical combinations of the basic dimensional quantities must be used. Examples include:

> **Density** is a measurement involving the mass (quantity) of a substance or material per unit of volume.

> **Pressure** is the amount of force compared to the surface area.

> **Volume** is space occupied, expressed in cubic units. It is a combination of three dimensions, including length, width, and height.

> **Speed** is the length (distance) of movement and time.

> **Acceleration** involves the mass of the object, a length (distance) of movement, and a unit of time of movement.

Errors

Since all measurements are made by senses or by extensions of senses, (instruments), and are actually approximations, measurements can not be expected to be "exact." Therefore, an error is expected. For example, a centimeter scale may be used to measure the length of an object. The actual length of the object may be slightly more or less than the centimeter increments on the scale. Therefore, a slight error may occur when the measurement is taken.

Percent Deviation (**percent error**) is obtained when mathematical calculations are used to solve problems involving measured and accepted values.

$$\text{percent error} = \frac{\text{difference from accepted value}}{\text{accepted value}} \times 100\%$$

For example, if a student using a balance determines the mass of an object to be 95 grams, but the mass of that object is actually 100 grams, there is an error. To determine the percent error, the student compares the observed mass and the actual mass by subtracting the measured mass from the actual mass:

100 grams	–	95 grams	=	5 grams
(actual mass)		(obtained mass)		(difference)

The percentage error is determined as follows:

5 grams divided by 100 grams x 100% = 5 % error

What are some characteristics of the properties of the environment?

Density is the concentration of the matter found within an object and is independent (does not depend) on the size and the shape of the material. Density is the mass of the object divided by the volume of the object, which is usually given in **g/cm³**:

$$\text{density} = \frac{\text{mass}}{\text{volume}}$$

The comparison of the density of different objects explains why some materials float on other objects. For example, a solid block of steel will sink when placed in a tub of water, but a ship made of steel plates will float in the ocean. This is due to **relative densities**.

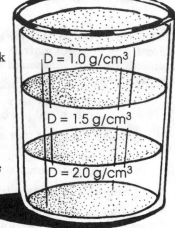

A solid cube of steel will sink, but if the relative density (mass/volume) of the steel is less than the density of the water, the steel will float (based on Archimedes' Principle: *a body immersed in fluid is buoyed up by a force equal to the weight of the fluid displaced*). The mass of the steel spread over a sufficient space (volume) will allow it to float (**flotation**).

Density Flotation

Phases Of Matter

Earth materials may exist in three main forms: solid, liquid, or gas. These three states of matter are dependent on the pressure or temperature conditions in which the material is placed. Generally, the material will change from a gas to a liquid to a solid as the temperature is lowered or as pressure is increased. As the temperature is increased or the pressure is lowered, the material will change state from a solid to a liquid to a gas.

There are several factors which affect the density, therefore the state, of a substance. Factors include:

1) *The density of a gas varies with pressure and temperature.* For example, in a weather system, warmer air rises through cooler air because the warmer air molecules are further apart (less dense). The cooler air is more compressed (air molecules are closer together, greater density, having higher pressure). Compared to the cool air, warm air has lower pressure. Therefore, the less dense warm air (rising) is replaced by the more dense cool air (sinking). Increasing the pressure on a gas causes the gas to contract (molecules move closer together). Decreasing the pressure produces the opposite effect. This principle also applies to liquids and solids, but to a lesser extent.

2) *The maximum density of most materials occurs in the solid phase.* For example, a granite rock will sink in a melt of granitic magma (liquid rock).

3) Water is a noted exception to the above. Water has its maximum density at a temperature of approximately 4°C. when it is a liquid. Therefore, the solid

form of water, ice, will float on liquid water. When water freezes to form ice, it expands. The water molecules move further apart, making ice less dense than liquid water.

Skill Assessments

Base your answers to questions 1 through 9 on your knowledge of Earth Science, the Reference Tables and the data in Tables I and II below. Tables I and II show the volume and mass of three samples of mineral *A* and three samples of mineral *B*.

Table I: Mineral *A*

Sample No.	Volume	Mass
1	2.0 cm^3	5.0 g
2	5.0 cm^3	12.5 g
3	10.0 cm^3	25.0 g

Table II: Mineral *B*

Sample No.	Volume	Mass
1	3.0 cm^3	12.0 g
2	5.0 cm^3	20.0 g
3	7.0 cm^3	28.0 g

Use the data to construct a graph on the grid provided at the right.

1   Mark an appropriate scale on the axis labeled "Mass (in grams)."
2   Plot a line graph for mineral A and label the line "Mineral *A*."
3   Plot a line graph for mineral B and label the line "Mineral *B*."
4   Write the formula for density:

5   Substitute the data for sample 3 of mineral *A* into the formula and determine the density of mineral *A*.

    Density of *A*:

MASS v. VOLUME

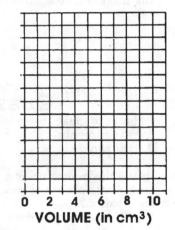

6   Substitute the data for sample 3 of mineral *B* into the formula and determine the density of mineral *B*.

    Density of *B*:

7   In one sentence tell what sample 2 of minerals *A* and *B* have in common.

8   In a sentence explain what happens to the masses of these minerals if their volume increases.

9   Explain what would happen to the density of sample 1 of mineral *B* if it is heated until it melts.

10   A person measures the length of a piece of wood to be 41 centimeters. If the actual length is 40. centimeters, what is the percent deviation (percent of error) from the actual length?

11   A student finds the mass of an igneous rock sample to be 48.0 grams. Its actual mass is 52.0 grams. What is the student's percent deviation?

12   Devise a classification system for the following rocks: rock salt, dolostone, rhyolite, sandstone, basalt, rock gypsum, conglomerate, shale, and granite.

In one sentence tell what property you used.

13   In one sentence, tell what property is used to classify the land-derived sedimentary rocks listed in the Earth Science Reference Tables.

14   A 5.00-milliliter sample of a substance has a mass of 12.5 grams. What is the mass of a 100.-milliliter sample of the same substance?

Questions For Topic I

1   In order to make observations, an observer must always use
   1   experiments   3   proportions
   2   the senses   4   mathematical calculations

2   In the classroom during a visual inspection of a rock, a student recorded four statements about the rock. Which statement about the rock is an observation?
   1   The rock formed deep in the Earth's interior.
   2   The rock cooled very rapidly.
   3   The rock dates from the Precambrian Era.
   4   The rock is black and shiny.

3   An interpretation based upon an observation is called
   1   a fact   3   a classification
   2   an inference   4   a measurement

4   A student observed a freshly dug hole in the ground and recorded statements about the sediments at the bottom of the hole. Which statement is an inference?
   1   The hole is 2 meters deep.
   2   Some of the particles are rounded.
   3   The sediments were deposited by a stream.
   4   Over 50% of the sediments are the size of sand grains or smaller

5   A predication of next winter's weather is an example of
    1   a measurement   3   an observation
    2   a classification   4   an inference

6   The graph at the right
    shows the amount of noise
    pollution caused by factory
    machinery during a
    one-week period. Which
    inference is best supported
    by the graph?

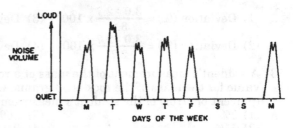

1   The machinery ran 24 hours a day.
2   The machinery was turned off on Saturday and Sunday.
3   The level of pollution remained constant during working hours.
4   The noise volume reached a peak on Friday.

7   A measurement is best defined as
    1   an inference made by using the human senses
    2   a direct comparison with a known standard
    3   an interpretation based on theory
    4   a group of inferred properties

8   Using a ruler to measure the length of a stick is an example of
    1   extending the sense of sight by using an instrument
    2   calculating the percent of error by using a proportion
    3   measuring the rate of change of the stick by making inferences
    4   predicting the length of the stick by guessing

9   The diagram at the right shows a rock and
    a standard mass on a double pan balance.
    Which statement about the rock is best
    supported by the diagram?
    1   The rock has a smaller volume than
        the standard mass.
    2   The rock has less mass than the
        standard mass.
    3   The rock has a greater force of gravity than the standard mass.
    4   The rock has a greater density than the standard mass.

10 A classification system is based on the use of
    1   human senses to observe properties of objects
    2   instruments to observe properties of objects
    3   observed properties to group objects with similar characteristics
    4   inferences to make observations

11 Which property was probably used to classify the substances in the chart at
    the right?
    1   chemical composition
    2   state (phase) of matter
    3   specific heat
    4   abundance within the Earth

Group A	Group B	Group C
water	aluminum	water vapor
gasoline	ice	air
alcohol	iron	oxygen

12 A rock's density is calculated as 2.7 g/cm³ but its accepted density is
    3.0 g/cm³. Which equation, when solved, will provide the correct percent
    deviation from the accepted value? [Refer to the Reference Tables]

(1) Deviation (%) = $\dfrac{3.0 - 2.7}{3.0}$ x 100   (3) Deviation (%) = $\dfrac{2.7}{3.0}$ x 100

(2) Deviation (%) = $\dfrac{3.0 - 2.7}{2.7}$ x 100   (4) Deviation (%) = $\dfrac{3.0}{2.7}$ x 100

13 A student's measurement of the mass of a rock is 30 grams. If the accepted
    value for the mass of the rock is 33 grams, what is the percent deviation
    (percent of error) of the student's measurement?
    (1) 9%   (3) 30%
    (2) 11%   (4) 91%

14 A student determines that the density of an aluminum sample is 2.9 grams
    per cubic centimeter. If the accepted value for the density of aluminum is 2.7
    grams per cubic centimeter, what is the student's approximate percent
    deviation?
    (1) 0.70%   (3) 7.4%
    (2) 0.20%   (4) 20%

15 Which factor can be predicted most accurately from day to day?
    1   chance of precipitation   3   chance of an earthquake occurring
    2   direction of the wind   4   altitude of the Sun at noon

16 The diameter through the equator of Jupiter is about 143,000 kilometers.
    What is this distance written in scientific notation (powers of 10)?
    (1) 143 x 10² km   (3) 1.43 x 10⁵ km
    (2) 1.43 x 10³ km   (4) 143 x 10⁵ km

17 In which phase (state) do most Earth materials have their greatest density?
    1   solid   2   liquid   3   gas

18 At which temperature does water have its greatest density?
    (1) –7°C   (3) 96°C
    (2) 0°C   (4) 4°C

19 The diagrams below represent two differently shaped blocks of ice floating in
    water. Which diagram most accurately shows the blocks of ice as they would
    actually float in water?

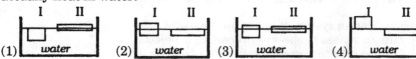

20 Compared to the density of water at room temperature, the density of ice is
    1   less   3   the same
    2   greater

21 If a wooden block were cut into eight identical pieces, the density of each
    piece compared to the density of the original block would be
    1   less   3   the same
    2   greater

22 As shown at the right, an empty
1,000.-milliliter container has a mass of
250.0 grams. When filled with a liquid, the
container and the liquid have a combined
mass of 1,300. grams.

What is the density of the liquid? [Refer to the
Reference Tables]
(1) 1.00 g/mL
(2) 1.05 g/mL
(3) 1.30 g/mL
(4) 0.95 g/mL

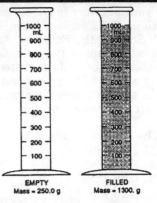

EMPTY
Mass = 250.0 g

FILLED
Mass = 1300. g

23 A mineral expands when heated. Which graph best represents the
relationship between change in density and change in temperature when
that mineral is heated?

(1) (2) (3) (4)

24 A 5.00-milliliter sample of a substance has a mass of 12.5 grams. What is
the mass of a 100.-milliliter sample of the same substance?
(1) 40.0 g   (2) 124 g   (3) 250. g   (4) 400. g

25 A student calculates the densities of five different pieces of aluminum, each
having a different volume. Which graph best represents this relationship?

(1) (2) (3) (4)

26 In which sequence are the earth layers arranged in order of increasing
average density?
1 atmosphere, lithosphere, hydrosphere
2 atmosphere, hydrosphere, lithosphere
3 hydrosphere, lithosphere, atmosphere
4 lithosphere, atmosphere, hydrosphere

27 Which graph best represents the average densities of the Sun, Moon, and
Earth?

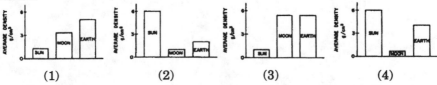

(1) (2) (3) (4)

Base your answers to questions to
questions 28-32 on your knowledge of
Earth Science, the Reference Tables
and the diagram at the right. Object *A*
is a solid cube of uniform material
having a mass of 65 grams and a
volume of 25 cubic centimeters. Cube *B*
is a part of cube *A*.

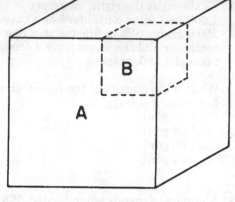

28 The density of cube *A* is
    (1) 3.8 g/cm³
    (2) 2.6 g/cm³
    (3) 0.38 g/cm³
    (4) 0.26 g/cm³

29 The density of the material in cube *A* is determined at different
    temperatures and phases of matter. At which temperature and in which
    phase of matter would the density of cube *A* most likely be greatest?
    [Assume a standard atmospheric pressure.]
    1   at 20°C and in the solid phase
    2   at 200°C and in the solid phase
    3   at 1800°C and in the liquid phase
    4   at 2700°C and in the gaseous phase

30 If cube *B* is removed from cube *A*, the density of the remaining part of cube
    A will
    1   decrease   2   increase   3   remain the same

31 The mass of cube *B* is measured in order to calculate its density. The cube
    has water on it while its mass is being measured. How would the calculated
    value for density compare with the actual density?
    1   The calculated density value would be less than the actual density.
    2   The calculated density value would be greater than the actual density.
    3   The calculated density values would be the same as the actual density.

32 If pressure is applied to cube *A* until its volume is one-half of its original
    volume, its new density will be
    1   one-half its original density   3   the same as its original density
    2   twice its original density   4   one-third its original density

The Changing Environment

Vocabulary To Be Understood In Topic II

Change	Equilibrium State	Interface
Cycles	Events	Pollution
Energy Flow	Frames of Reference	Rate Changes

A. Nature Of Change

How can changes be described?

Characteristics Of Change

The Earth environment is in a constant state of change. Although the total material mass of the Earth remains relatively constant, it is in a changing state. The law of the conservation of matter and energy states that *under normal conditions matter can neither be created nor destroyed – but, can change form.* Change in an Earth system or an object can be described as the occurrence of an **event**. This event (change) may occur suddenly (such as lightning during a thunderstorm). Change may also be observed to take very long periods of time (for example, the formation of a mountain).

Frames Of Reference

To understand and classify Earth changes, two primary frames of reference used are **time** and **space**. Geologic history can be described and dated according to ages and eras. This allows for a chronological record of Earth changes based on relative age evidence and dating techniques. Seasons of the year are characterized by time and events. The positions and phases of the Moon provide frames of reference for fishermen (tides) and farmers (planting and harvesting).

Rate Of Change

Many events or changes in the Earth can be best described by reference to the rates of the change. Some changes take very little time, as in weather systems, which may take only hours or days to occur, or in the complete phase changes of the Moon (weeks). Other changes may take such long periods of time that it becomes very difficult to accurately measure these changes, such as the movement of the Earth plates or the erosion of a mountain range.

Cycles – Non-cycles

Most changes in the environment are cyclic. **Cyclic** changes involve events that repeat in time and space, usually in an orderly manner. For example:

1) Earth materials are cycled and recycled through definite patterns, such as the water cycle.

2) Through alternating changes in state (phase) (e.g., evaporation and condensation), water is cycled and purified between the ground, water reservoirs, and the atmosphere.

3) Seasonal changes, such as freezing and thawing, produce predictable changes such as the primary type of weathering that is dominant.

4) The Sun, Moon, stars, planets, and other celestial objects have definite cyclic motions.

General weather patterns repeat in an orderly manner. However, not all changes are cyclic. There are "one-direction" events, such as the radioactive decay of certain elements, the extinction of a species, or the impact of a meteorite on the Moon.

Predictability Of Change

If there is *sufficient evidence and knowledge* of the nature of the environmental change, it may be possible to predict the scope (type and amount) and direction that future changes will take. Since changes in the regular cycles of some celestial objects occur very slowly, it is possible to predict when and where a comet will appear, or when and where the Sun's rays will strike the Earth's surface.

Meteorologists are able to predict general storm tracks or forecast hurricanes and major weather disturbances because of the many years of collecting data on weather. It should be noted that *general events are fairly predictable (such as the sunrise/sunset), whereas individual occurrences are much more difficult to precisely predict (such as the weather).*

Factors involved in change are called **variables**. The more observations and the fewer variables involved, the greater the accuracy of the prediction of the change.

B. Energy And Change
What is the relationship of energy to change?

Energy Flow And Exchange

When environmental change occurs, energy is lost by one part (*source*) of the environment and gained in another part (*sink*). (See Topic V, Section C for further explanation.) This change occurs **simultaneously** (at the same time). For example, as the energy of the Sun strikes the Earth's crust and is absorbed, the radiant energy produces heat to warm the surface of the Earth. The energy within the high speed winds of a hurricane or tornado is absorbed by trees, buildings, and other things, causing destruction and erosion.

This *exchange of energy (loss and gain)* occurs at an **interface** (location or boundary) between the affected parts of the environment. Change in the environment occurs at an interface between materials, such as air to rock. For example, unless there is an interface *between* the hurricane winds and objects, there is virtually no destruction. However, over the ocean, the interface of the winds of the hurricane and the environment is the water surface, which may absorb much of the energy and transmit the rest in the form of increased wave activity. Should these waves themselves interface with land, buildings, or boats, a second interface occurs, continuing the energy change.

C. Environmental Change
How do humans modify the environment?

Environmental Balance

Our environment is in a **state of equilibrium** that tends to remain unchanged because, when equilibrium is disturbed nature works to establish a new one. *Change is also constant*, so what appears to be unchanged, may indeed be slowly changing. For example, from week to week or even year to year, a stream may appear to be unchanged. However, over long periods of time (decades or centuries) the stream may change its course (meander). The flora (plants), fauna (animals), mineral content, and stream banks may change as well.

Should there be a sudden change in the climatic or weather conditions, such as severe rainfall causing flooding, there could be a vast change in the stream. Erosion could change the stream's banks, its course, and its contents.

Because of modern technology, humans are capable of rapid and large disruption of the environmental equilibrium, either in positive conservation or in destruction. Through the building of dams, roadways, and communities, people have irreversibly destroyed millions of acres of forests and natural wetlands.

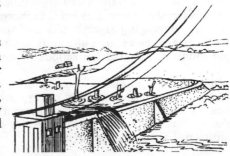

On the other hand, technology has enabled humans to preserve many of nature's wonders. For example, major erosion of the American side of Niagara Falls during the first part of this century endangered the very survival of the falls. By changing the direction and flow of the water and by building unseen reinforcement in the cliffs, human engineering has been able to insure the continuance of the American Falls.

Environmental Pollution

Environmental pollution has reached serious levels in recent years, due to human neglect of the environmental equilibrium. The environment is considered polluted when the concentration of *any substance or form of energy* reaches a proportion that adversely affects humans, plants, and/or animals or the environment on which all living things depend.

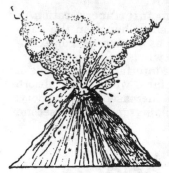

Environmental pollutants are the result of both natural environmental disruptions and the technological oversights of humans (such as activities of individuals and industrial processes). The eruption of Mt. Saint Helens in the state of Washington, was an example of natural pollution. It produced millions of tons of ash and destroyed thousands of acres of plant and animal life. In addition, the airborne particles (such as dust) from the eruptions caused climatic and weather changes all across North America, affecting farm crops and air quality. Another example of a common natural pollutant is pollen.

People appear to have done much more damage than natural pollutants have done to the environmental equilibrium. This has been through the excessive addition of pollutants to the environment. Pollutants include such diverse materials as solids, liquids, gases, biologic organisms, and forms of energy such as heat, sound, and nuclear radiation and wastes.

Both natural processes and human pollution tend to vary with seasons, days of the week, and times of days. In nature, rivers tend to **purge** (churn up from the bottom) accumulated wastes each late summer or early fall. This is a "cleaning" process for the rivers, but leads to increased surface pollution, disruption of wildlife, and damaging river banks and water supplies. During weekdays and high traffic times, more air pollutants are added to the atmosphere, producing smog and industrial **toxins** (poisonous waste chemicals).

Because the summer usually brings about an increase in electrical needs, there is more pollution from energy producing plants. Excess heat discharge increases aerobic and anaerobic bacteria, causing a decrease in plant and animal life. In the winter, a greater need for heat in buildings increases the output of additional air pollutants.

Skill Assessments

Base your answers to questions 1 through 6 on your knowledge of Earth Science, the Reference Tables, and the graph and data below. The data table shows the concentration of air pollutants per cubic centimeter for two days over an eastern city in the United States.

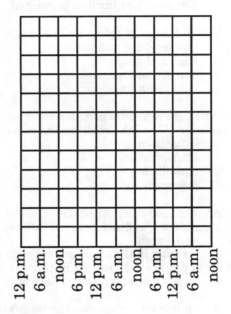

Data Table		
	Time	Poll/cm3
Monday, May 1st	12 p.m. 20,000	
	6 a.m. 40,000	
	noon 30,000	
	6 p.m. 60,000	
	12 p.m. 38,000	
Tuesday, May 2nd	6 a.m. 50,000	
	noon 41,000	
	6 p.m. 68,000	
	12 p.m.. 40,000	

1   On the vertical axis make an approximate scale for pollutants in parts/cm^3.

2   Using the data from the table, plot the points and construct a line graph of pollutants/cm^3.

3   Is the graph cyclic or noncyclic? Why?

4   In one sentence, compare the concentration of pollutants at 6 p.m. on Monday with the concentration of pollutants at 6 p.m. on Tuesday.

5   In a sentence or two, explain the likely cause of the pollution peaks.

6   Predict the time of the next pollution peak. Predict the concentration of pollutants.

Questions For Topic II

1   What happens when a change occurs?
1   Pollution is produced.
2   The temperature of a system increases.
3   The properties of a system are altered.
4   Dynamic equilibrium is reached

2   Which frames of reference are best to use to describe all Earth changes?
1   temperature and pressure
2   time and space
3   density and mass
4   volume and weight

3   When the amounts of water entering and leaving a lake are balanced, the volume of the lake remains the same. This balance is called
1   saturation
2   transpiration
3   equilibrium
4   permeability

4   An interface can best be described as
1   a zone of contact between different substances across which energy is exchanged
2   a region in the environment with unchanging properties
3   a process that results in changes in the environment
4   a region beneath the surface of the Earth where change is not occurring

5   Which line best identifies the interface between the lithosphere and the troposphere?

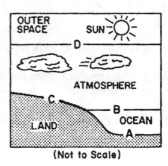
(Not to Scale)

1   line A
2   line B
3   line C
4   line D

6   Future changes in the environment can best be predicted from data that are
1   highly variable and collected over short period of time
2   highly variable and collected over long periods of time
3   cyclic and collected over short periods of time
4   cyclic and collected over long periods of time

7   Which statement best explains why some cyclic earth changes may *not* appear to be cyclic?
1   Most Earth changes are caused by human activities.
2   Most Earth changes are caused by the occurrence of a major catastrophe.
3   Most Earth changes occur over such a long period of time that they are difficult to measure.
4   No Earth changes can be observed because the Earth is always in equilibrium.

8   Over several years, the apparent size of the Sun as viewed by an observer on
     Earth will probably
     1   vary in a cyclic manner   3   increase at a regular rate
     2   decrease at a regular rate   4   vary in an unpredictable manner

9   In which Earth process is the rate of change easiest to measure?
     1   erosion of a mountain   3   formation of a rock
     2   discharge of a stream   4   development of a mature soil

10   During a ten year period, which is a noncyclic change?
     1   the Moon's phases as seen from the Earth
     2   the Earth's orbital velocity around the Sun
     3   the impact of a meteorite on the Earth
     4   the apparent path of the Sun as seen from the Earth

11   When the amounts of biologic organisms, sound, and radiation added to the
     environment reach a level that harms people, these factors are referred to as
     environmental
     1   interfaces   3   phase changes
     2   pollutants   4   equilibrium exchanges

12   Which graph best represents the typical relationship between population
     density near a lake and pollution of the lake?

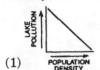   (1)   (2)   (3)   (4)

13   Which diagram illustrates the process that best cleans the atmosphere?

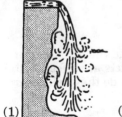   (1)      (2)   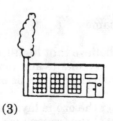   (3)      (4)

14   Air pollution from industrial processes may be most effectively reduced by
     1   growing more green plants to absorb the pollutants
     2   adding anti-irritants to the air
     3   removing waste materials before they enter the atmosphere
     4   precipitating the pollutants by heating the air

15   Which is the least probable source of atmospheric pollution in heavily
     populated cities?
     1   human activities
     2   industrial plants
     3   natural processes
     4   automobile traffic

16 The map at the right illustrates the distribution of acid rain over the United States on a particular day. The isolines represent acidity measured in pH units.

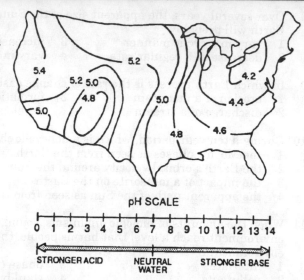

According to the pH scale shown below the map, which region of the United States has the greatest acid rain problem?
1 northeast
2 northwest
3 southeast
4 southwest

17 Some scientists predict that the increase in atmospheric carbon dioxide will cause a worldwide increase in temperature. Which could result from this increase in temperature?
1 Continental drift will increase.
2 Land masses will shift toward the Equator.
3 Additional land masses will form.
4 Ice caps at the Earth's poles will melt.

18 As the concentration of pollution particles in the atmosphere increases, the amount of insolation that reaches the Earth's surface will most likely
1 decrease
2 increase
3 remain the same

19 Some scientists believe that high-flying airplanes and the discharge of fluorocarbons from coolants and making foam products are affecting the atmosphere. Which characteristic of the atmosphere do they believe is affected?
1 composition of the ozone layer of the stratosphere
2 wind velocity of the tropopause
3 location of continental polar highs
4 air movement in the doldrums

The Earth Model — Measuring The Earth

Vocabulary To Be Understood In Topic III

Altitude	Gradient	Model of Earth
Atmosphere	Gravitational Force	Oblate Spheroid
Contour Line	Hydrosphere	Parallels
Contour Map	Isoline & Iso-surface	Polaris (North Star)
Coordinate System	Latitude	Prime Meridian
Field	Lithosphere	Scalar Quantity
Geographic Poles	Longitude	Vector Quantity

A. Earth Dimensions

How can the Earth's shape be determined?

Shape Of The Earth

The Earth appears to be the shape of a sphere (round in circumference), when observed from space or scaled down to a model, such as a classroom globe. However, by actual measurement the Earth is not a perfect sphere. Instead, it has a larger circumference at the equator (0° latitude), than through the poles (0° longitude).

The circumference of the Earth at the equator is 40,076 kilometers, and the equatorial diameter is 12,757 kilometers. The circumference of the Earth through the poles is 40,008 kilometers, and the polar diameter is 12,714 kilometers. Therefore, the Earth is slightly "bulged" at the equator and slightly "flattened" at the poles. Thus, the true shape of the Earth is best defined as an **oblate spheroid**.

Evidence For The Earth's Shape

Observations of the North Star (Polaris). If a straight line passed from the South Pole, through the center of the Earth (polar axis) to the North Pole, and continued into space, it would very nearly pass through Polaris as well. Therefore, Polaris is termed the North Star, as it lies in space practically over the geographic north pole of the Earth. In order to understand why Polaris can be used to help determine the Earth's shape, it is necessary to understand the "geometry of a sphere" and the terms latitude and altitude.

When observing an object in the atmosphere or space, the object's **altitude** refers to its angle (measured in degrees) above the horizon. When locating a point on the Earth's surface, the term **latitude** describes the point's position (measured as an angle in degrees) north or south of the Earth's equator.

According to a principle of the "geometry of a sphere," as a sphere is rotated the angle of a fixed point outside of a sphere, compared to the surface of that sphere, is equal to the the angle of the sphere's rotation. Therefore, if a sphere is rotated 45°, a fixed point outside of that sphere will be at an angle of 45° from its original point, before the sphere's rotation. (Note the illustration below.)

Considering Polaris to be a fixed point above the North Pole, it's angle (altitude) to the Earth's polar axis should be the same as the angle of the Earth's tilt on that same axis. *Indeed, the altitude of Polaris corresponds very closely with an observer's latitude.* But, when the angles are measured with precise instruments, the accurately measured altitude of Polaris is *not exactly* the same as the observer's latitude on Earth. Therefore, Polaris gives evidence that the Earth is not a *perfect* sphere; instead, it is *slightly* out of round, or **oblate**.

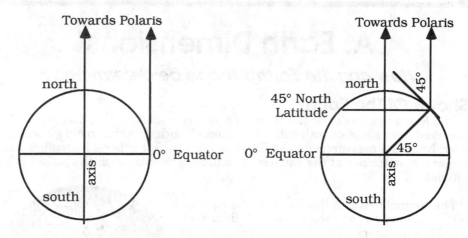

As seen in these illustrations, the altitude of Polaris should be the same as the observer's latitude on the Earth, if the Earth is a perfect sphere.

Photographs of the Earth from space. With the exploration of space, came the ability to take very precise photographs of the Earth from great distances in space. When precisely measured, these photographs show the Earth to be larger at the equator and flatter at the geographic poles. However, *the shape of the Earth, when drawn to scale on a sheet of paper, appears to be perfectly round.*

Gravimetric (gravity) measurements. Gravity is the force of attraction between any two objects. Since the Earth has such a large mass, smaller objects with less mass are attracted (pulled) towards the Earth. If the Earth were a perfect sphere, it would be expected to exert an equal pull on objects anywhere on or above the Earth's surface at equal distances from the center of the Earth.

The law of gravitation states that a gravitational force is proportional to the square of the distance between the two centers of attracted objects. In other words, the weight of an object moved anywhere on the surface of the Earth (assuming Earth is a perfect sphere), should remain the same. However, precise measurements of objects indicate that the same object weighs more at the poles than at the equator of the Earth. Even accounting for the centrifugal forces produced by the Earth's rotation, there is still a greater gravitational pull at the poles than at the equator. (Note, the Earth's oblateness in the below left illustration is greatly exaggerated.)

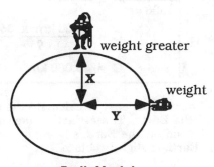

Earth Model
Y greater distance than X,
therefore, X has greater
force of gravity.

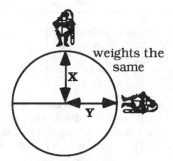

Sphere Model
Y and X are equal distances,
therefore, Y and X have
equal forces of gravity.

How can the Earth's size be determined?

Measuring The Earth's Size

All of the dimensions of the Earth can be determined from observations of the Earth from space. The relative positions of the Earth and the Sun can be used to determine the exact size of the Earth.

Before the space age began, even as far back in history as the Greeks, fairly good estimates of the Earth's dimensions could be made. For example, by measuring the altitude of the Sun at two different places on the Earth's surface at exactly the same time of day, the Earth's circumference can be mathematically determined.

For example: Two observers are standing on the same meridian, a distance of 1000 kilometers apart. On the same day at the same time, each measures the altitude of the Sun. Finding the difference between the two altitudes of the Sun gives the angle at the center of the Earth that separates the two observers. Assuming that the Earth is a perfect sphere with 360°, dividing the angle difference between the two observers into the 360° and multiplying that number by the 1000 kilometer distance between them, gives the total circumference of the Earth.

Determining The Circumference Of The Earth

9° divided into 360° = 40° ($^1/_{40}$ of 360° circumference)
1000 km multiplied by 40 = 40,000 km (Earth's circumference)

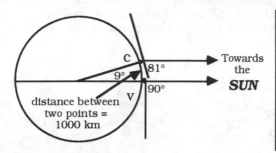

Determining by Proportion:
The Circumference Of The Earth

$$\frac{9°}{1000 \text{ km}} = \frac{360°}{C}$$

$$C = \frac{1000 \text{ km} \times 360°}{9°}$$

$$C = 40,000 \text{ km}$$

When the circumference of the Earth is known, it is easy to calculate the surface area, radius, diameter, and volume of the Earth. These Earth dimensions are not exact because these calculations depend on the Earth's being a perfect sphere. However, we have learned that the Earth is slightly oblate.

The following formulas are used to calculate the dimensions of the Earth:

Where:
- **C** is the circumference
- **r** is the radius
- **D** is the diameter
- **V** is the volume
- **A** is the surface area

$$r = \frac{C}{2\pi} \qquad D = 2r$$

$$V = \frac{4}{3}\pi r^3 \qquad A = 4\pi r^2$$

What is the extent of the atmosphere, hydrosphere, and lithosphere?

The Earth's Atmosphere

The atmosphere of the Earth is composed of the materials (solids, liquids, and gases) which form a thin envelope surrounding the Earth, held in place by gravitation and rotation with the Earth. The atmosphere extends several hundred kilometers above the Earth's surface into space and is the least dense of the Earth's three spheres.

The atmosphere is **stratified** (layered) into **zones**, each having distinct characteristics, including temperature and pressure ranges, composition, and effects produced on the Earth's surface. These zones include (from the Earth's surface towards space):

Troposphere,
Stratosphere,
Mesosphere, and
Thermosphere.

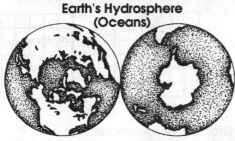

**Earth's Hydrosphere
(Oceans)**

North Pole View   **South Pole View**

Earth's Hydrosphere

About four-fifths (±70%) of the Earth's surface is covered with a relatively thin film of water. Compared to the other zones of the Earth, it is very thin, only averaging 3.5 to 4.0 kilometers thick, much like the skin on an apple.

The hydrosphere includes all bodies of water, marine (salt-containing oceans) and fresh (inland lakes and rivers)

Earth's Lithosphere

The most solid portion of the Earth is the rock near the Earth's surface and is a continuous solid shell, often under the hydrosphere, It accounts for the general underwater features of the Earth, including mountains, valleys, and the ocean floor.

B. Positions On The Earth

*How can a position
on the Earth's surface be
determined?*

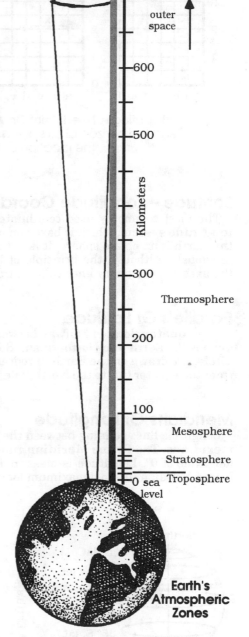

Earth's
Atmospheric
Zones

Coordinate Systems

There are many reference systems that can be used to determine positions on the Earth. A coordinate system uses a **grid** of imaginary lines and two points, called **coordinates**, to locate a particular position on the surface of the Earth. A fixed point can be located on a graph by identifying the intersecting point of two lines (axes).

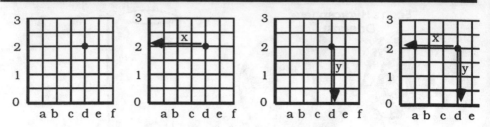

Locating A Fixed Point On A Graph (Coordinate System)
1st - Read the horizontal axis **x = d**   2nd - Read the vertical axis **y = 2**
The location of the point is at **d,2**.

Latitude – Longitude Coordinate System

The most commonly used coordinate system for the Earth is the **latitude – longitude system**, which is based on celestial observations or star angles. Since the Earth is an oblate sphere, it is not practical to use a flat graph as illustrated previously. Although the principle of locating a fixed point remains the same, the axes are imaginary lines called parallels and meridians circling the Earth.

Parallels Of Latitude

The **equator**, located halfway between the geographic poles, is a circle which divides the Earth into Northern and Southern Hemispheres. Circles, called **parallels**, are drawn on the Earth to represent latitude. These latitude circles range from the equator (0°) to the North Pole (90° N) and to the South Pole (90° S).

Meridians Of Longitude

Longitude lines running between the North Pole and the South Pole are called **meridians**. The **Prime Meridian** runs through Greenwich, England and has a longitude of 0°. Longitude is measured using the meridians east or west of the Prime Meridian (0°) to a maximum longitude of 180°.

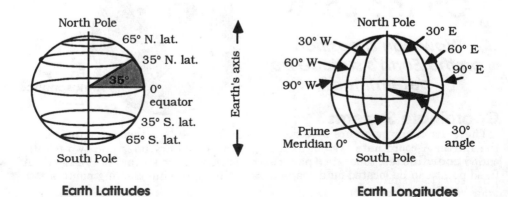

Earth Latitudes   **Earth Longitudes**

Latitude And Longitude Measurements

Both latitude and longitude are measured in degrees. Latitude is distance measured in degrees north or south of the equator (0° to 90° North, 0° to 90° South). Longitude is distance measured in degrees east or west of the Prime Meridian (0° to the 180° meridian).

Latitude. The altitude of Polaris above the horizon is often used to determine a particular latitudinal position in the Northern Hemisphere (refer to Topic III, Section A). In the Southern Hemisphere other stars are used. Since travel in space, in air, and on the water require very precise navigation, astronomical tables are also used. They account for variations in the positions of heavenly bodies based on time and date.

Longitude. The Earth rotates from west to east, 360° in a 24 hour period of time, or a distance of 15° of longitude per hour. Comparing your local time with the time at Greenwich, England (0°), enables you to determine your longitude. For every hour difference in time between an observer and Greenwich Mean Time (GMT), the observer is 15° away from the Prime Meridian.

If the Sun is on the Prime Meridian, it is 12 noon at Greenwich, England. Since the Earth rotates west to east, the Sun has not yet reached it highest point (noon time) to the west of Greenwich; therefore, time is earlier to the west. However, to the east of Greenwich, the Sun has already passed its highest point; therefore, time is later to the east of Greenwich.

For example, a ship sailing in the Atlantic Ocean maintains a clock (chronometer) with Greenwich Mean Time. If a time comparison is taken when the Sun is at its highest point over the ship (the ship's "local noon") and found to be 3 hours earlier than Greenwich Mean Time; than we know that the ship is at 45° West Longitude, since the ship's local time is 3 hours earlier than the same time in Greenwich (3 x 15° = 45°). Modern navigational techniques employ the use of radio signals from Earth and satellite bases.

How can the characteristics of a position be measured and described?

Vector And Scalar Quantities

Fields

A field is a region of space that contains a measurable quantity at every point. Other examples of fields include: gravity, temperatures, atmospheric pressures, and gradients (changes in elevations). Field quantities may be vector and scalar.

A **vector quantity** or **field** is defined as having both magnitude and direction. For example, magnetism and gravity are vector quantities. A **scalar quantity** or **field** has a magnitude or size only. For example, the atmospheric temperature or the relative humidity are scalar quantities.

Isolines

In order to visualize vector or scalar field quantities, maps can be drawn by using isolines. Maps containing isolines are **models** representing field characteristics in two dimensions. Isolines connect points that are equal (or the same) in value. For example, a comparison of the atmospheric pressure over a certain area is shown with **isobars**, elevations on a relief map are shown with **contour lines**, and lines connecting equal temperatures on a weather map are called **isotherms**.

Iso-surfaces

Isolines can only show quantities on a two-dimensional surface, but iso-surfaces are models representing field characteristics in three dimensions, such as a magnetic field or a contour map of varying elevations, temperatures or pressures. Note that all of the points of an iso-surface must have the same field value.

Field Changes

Fields are generally not **static** (unchanging). The characteristics of fields frequently change with the passage of time.

For example, the temperatures and air pressures over the surface of the Earth are in a constant state of change. Therefore, a weather map shows only those field conditions occurring during a specific time and date. Also, a contour map made several years ago of a particular landscape would have to be updated from time to time to remain accurate, due to the changes produced by the forces of nature and humankind.

Gradient

A gradient or average slope within the field expresses the rate of change of the field quantity from one place to another place. The following formula is used to determine the rate of change:

$$\text{gradient} = \frac{\text{the amount of change in the field}}{\text{the change of distance in the field}}$$

Example:
If a weather map shows a change in atmospheric pressure from 996 mb to 1004 mb between towns that are 400 kilometers apart, then the rate of change (gradient) is 8 mb per 400 kilometers or 2 mb/100 kilometers.

(Note. Gradient on the east side of the LOW is *steep*; whereas, the gradient on the west side is *gentle*.)

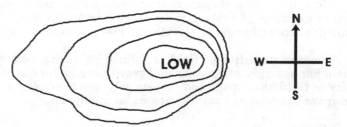

The Contour (Topographic) Map

The topographic map uses **isolines**, called **contour lines**, to connect points of same elevation (usually based on sea level measurements). Generally, every 5th line is an *index contour line*, which means that it is printed darker and is interrupted to give the elevation.

The difference in elevation between two consecutive contour lines is the **contour interval**. Common intervals are 5, 10, 20, 50 and 100 feet. Enclosed depressions are shown with hachured contour lines (the hachure marks ⁄⁄⁄⁄⁄⁄ pointing down-slope).

The **map legend** gives **distance scales** along with a **key** of symbols representing natural and cultural (man-made) features of the field, such as cities, roads, buildings, marshes, and forests.

The following illustration represents a typical topographic map with questions and a profile question. (Note. When crossing a stream or valley, contour lines bend in a "V" shape pointing uphill.)

Skill Assessments

Base your answers to questions 1 through 5 on the diagram at the right represents a temperature field map of the air near the ceiling of a room. The letters represent points in the field.

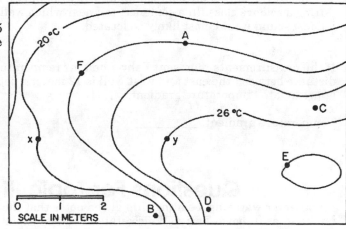

1　What kind of isolines are shown on this field map?

2　In a sentence, tell between which two letters the greatest temperature gradient is found.

3　Write the equation for determining gradient.

4　Substitute the appropriate data into the equation and determine the gradient between letters *x* and *y*.

5　In a sentence, tell what happens to the temperature along the isoline from *A* to *F*.

Base your answers to
questions 6 through 9
on the diagram at the
right that represents
field-intensity
measurements taken at
equal elevations within
a room. Letters *A*,*B*,
and *C* are reference
points on the plane
where the readings
were taken.

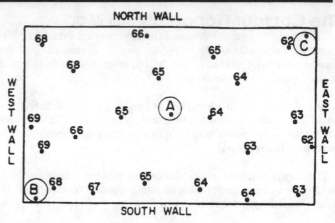

6   Estimate the field-intensity value at point *A*.

Point *A* value: _____

7   Draw in the isolines to make a field map using an interval of 1.

8   The measurements represent sound intensity (loudness) readings taken
during a science class. In a sentence, tell near which wall the source of the
loudest sound would most likely be located.

9   If the measurements represent Fahrenheit air temperatures, and the
distance between the east and west wall is 12 meters, determine the
approximate temperature gradient.

Temperature gradient: _____

Questions For Topic III

1   An observer watching a sailing ship at sea notes that the ship appears to be
"sinking" as it moves away. Which statement best explains this
observation?
   1   The surface of the ocean has depressions.
   2   The Earth has a curved surface.
   3   The is is rotating.
   4   The Earth is revolving.

2   The true shape of the Earth is best described as a
   1   perfect sphere   3   slightly oblate sphere
   2   perfect ellipse   4   highly eccentric ellipse

3   According to the Reference Tables, the radius of the Earth is approximately
   1   637 km   3   63,700 km
   2   6,370 km   4   637,000 km

4   How many degrees per hour must a satellite orbit in order to remain above
    the same spot on the Equator?
    1   1° per hour   3   24° per hour
    2   15° per hour   4   360° per hour

5   At sea level, which location would be closest to the center of the Earth?
    1   45° South latitude   3   23° North latitude
    2   the Equator   4   the North Pole

6   At which location would an observer find the greatest force due to the
    Earth's gravity?
    1   North Pole   3   Tropic of Cancer (23½°N.)
    2   New York State   4   Equator

7   Which statement most accurately describes the Earth's atmosphere?
    1   The atmosphere is layered, with each layer possessing distinct
        characteristics.
    2   The atmosphere is a shell of gases surrounding most of the Earth.
    3   The atmosphere's altitude is less than the depth of the ocean.
    4   The atmosphere is more dense than the hydrosphere but less dense than
        the lithosphere.

8   The water sphere of the Earth is known as the
    1   atmosphere   2   hydrosphere 3   lithosphere   4   troposphere

9   Based on the diagram at the right,
    what is the circumference of
    planet *Y*?
    (1) 9,000 km
    (2) 12,000 km
    (3) 16,000 km
    (4) 24,000 km

10 Which statement provides the best evidence that the Earth has a nearly
    spherical shape?
    1   The Sun has a spherical shape.
    2   The altitude of Polaris changes in a definite pattern as an observer's
        latitude changes.
    3   Star trails photographed over a period of time show a circular path.
    4   The length of noontime shadows change throughout the year.

11 The polar circumference of the Earth is 40,008 kilometers. What is the
    equatorial circumference?
    (1) 12,740 km   (2) 25,000 km   (3) 40,008 km   (4) 40,076 km

12 Which diagram most accurately shows the cross-sectional shape of the Earth
    drawn to scale?

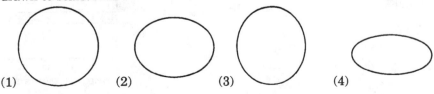

    (1)   (2)   (3)   (4)

13 The latitude of a point in the Northern Hemisphere may be determined by measuring the
    1   apparent diameter of Polaris   3   distance to the Sun
    2   altitude of Polaris   4   apparent diameter of the Sun

14 An observer in New York State measures the altitude of Polaris to be 44°. According to the Reference Tables, the location of the observer is nearest to
    1   Watertown   2   Elmira   3   Buffalo   4   Kingston

15 An observer travels northward from New York State, the altitude of the North Star
    1   increases directly with the latitude
    2   decreases directly with the latitude
    3   increases directly with the longitude
    4   decreases directly with the longitude

16 If the deepest parts of the ocean are about 10 kilometers and the radius of the Earth is about 6,400 kilometers, the depth of the ocean would represent what percent of the Earth's radius?
    1   less than 1%   3   about 25%
    2   about 5%   4   more than 75%

17 The Earth's latitude system is based upon measurements of
    1   star angles   3   magnetic directions
    2   gravity intensities   4   geographic landmarks

18 According to the Reference Tables, which city is located closest to 44°N latitude, 76° W longitude?
    1   Massena   2   Binghamton   3   Buffalo   4   Watertown

19 As a ship crosses the Prime Meridian, the altitude of Polaris is 65°. What is the location of the ship?
    1   65° South latitude, 0° longitude
    2   65° North latitude, 0° longitude
    3   0° latitude, 65° West longitude
    4   0° latitude, 65° East longitude

20 The graph at the right shows the percentage distribution of the Earth's surface elevation above and depth below sea level.

Approximately what total percentage of the Earth's surface is below sea level?
    (1) 30%
    (2) 50%
    (3) 70%
    (4) 90%

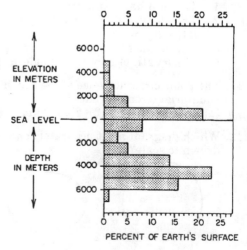

PERCENT OF EARTH'S SURFACE

Base your answers to questions
21 and 22 on the diagram which
represents a contour map of a
hill.

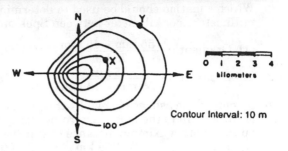

Contour Interval: 10 m

21 On which side of the hill does
the land have the steepest
slope?
1   north   3   east
2   south   4   west

22 What is the approximate gradient of the hill between points *X* and *Y*?
1   1 m/km   3   3 m/km
2   10 m/km   4   30 m/km

23 What is the elevation of the highest
contour line shown on the map?

1   10,000 feet
2   10,688 feet
3   10,700 feet
4   10,788 feet

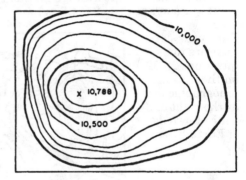

Base your answers
to questions 24-26
on the outline map
of the state of
Washington, the
Reference Tables
and your
knowledge of
Earth Science. The
map shows isolines
of ashfall depths
covering a portion
of the State of
Washington which
resulted from a
volcanic eruption
of Mount St.
Helens.

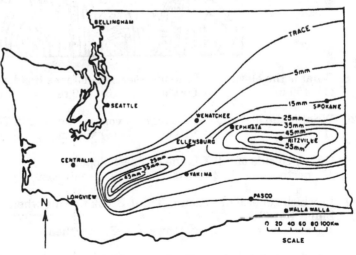

24 Which two towns received the same depth of volcanic ash?
1   Pasco and Wenatchee
2   Yakima and Ephrata
3   Centralia and Ritzville
4   Spokane and Ellensburg

25 Which equation should be used to determine the ashfall gradient in millimeters per kilometer between Spokane and Ritzville?

(1) $\text{Gradient} = \dfrac{55 \text{ mm} - 15 \text{ mm}}{75 \text{ km}}$ (3) $\text{Gradient} = \dfrac{55 \text{ mm} - 15 \text{ mm}}{100 \text{ km}}$

(2) $\text{Gradient} = \dfrac{75 \text{ km}}{55 \text{ mm} - 15 \text{ mm}}$ (4) $\text{Gradient} = \dfrac{100 \text{ km}}{55 \text{ mm} - 15 \text{ mm}}$

26 If some of the ash was blown by the prevailing wind from the western edge of the ashfall to the northeastern border of Washington, what was the approximate maximum distance that it traveled?
(1) 150 km   (2) 250 km   (3) 500 km   (4) 740 km

27 A stream has a source at an elevation of 1,000 meters. It ends in a lake that has an elevation of 100 meters. If the lake is 200 kilometers away from the source, what is the average gradient of the stream?
1   1.5 m/km   2   4.5 m/km   3   10 m/km   4   15 m/km

Base your answers to questions 28-32 on the topographic map at the right and your knowledge of Earth Science.

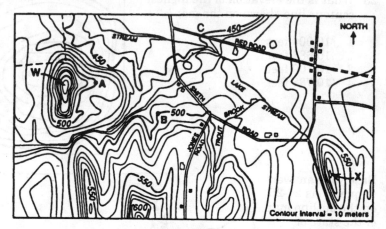

28 What is the elevation of the intersection of Jones Road and Smith Road?
(1) 450 m   (2) 500 m   (3) 550 m   (4) 600m

29 What is the elevation of the highest contour line of hill W?
(1) 440 m   (2) 510 m   (3) 560 m   (4) 610 m

30 On which side of hill X is the steepest slope?
1   north   2   east   3   southeast   4   southwest

31 In which general direction is Trout Brook flowing when it passes under Smith Road?
1   northeast   2   northwest   3   southeast   4   southwest

32 Which diagram best represents the profile along a straight line between points A and B?

  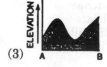 

The Earth Model — Earth Motions

Vocabulary to be Understood in Topic IV

Angular Diameter
Aphelion, Perihelion
Apparent Daily Motion
Apparent Solar Day
Celestial Objects, Model
Centrifugal, Coriolis Effects
Cyclic Energy Transformation
Elliptical Eccentricity, Focus
Equinox, Solstice
Foucault Pendulum
Geo- , Heliocentric Models
Gravitation

High Noon
Kinetic, Potential Energy
Mean Solar Day
Orbit, Arc
Orbital Speed (Velocity)
Phases of the Moon
Revolution, Rotation
Satellite
Solar System
Terrestrial Motions
Time - Day, Year
Zenith Point

A. Celestial Observations

What observations can be made of celestial objects?

The Motions Of Objects In The Sky

Collectively, the objects observed in the sky during the day or the night are called **celestial objects**. These objects include the other planets, our Sun and Moon, the stars, and comets. For the most part, these celestial objects have an **apparent circular motion from east to west** in the sky as observed from the Earth's surface. This motion appears as a part of a circle around the Earth, called an **arc**.

The apparent center of the arcs of the stars is very near to Polaris. This apparent circular motion causes the celestial objects to appear to rise in the east and set in the west.

Actually, the **apparent daily motion** of the stars, Sun, planets, and Moon is due to the Earth's rotation, at a rate of 15° per hour or one complete Earth rotation (360°) in 24 hours (see Topic III, Section B).

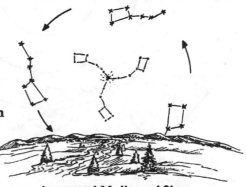

Apparent Motion of Stars

Apparent Circular Motion Models

There are two primary models to explain the motions of celestial objects. The **geocentric model** shows a stationary Earth as the center of the celestial motions. This model was believed to be correct from the time of the ancient Greek astronomers until the 16th century. Since then, the most commonly accepted model for celestial motions is the **heliocentric model**. This model shows the Sun as the center of planetary motion with the Earth rotating on its axis as it revolves around the Sun. It takes about 365¼ Earth days (24 hrs each) for the Earth to make one complete revolution around the Sun.

A **revolution** is defined as the movement of one body around another body. This motion is also referred to as orbiting. An **orbit** is the path taken by one body as it revolves around the other body. In this case the Earth "orbits" the Sun. The **rotation** of a body is the spinning (turning) of the body on its axis. One complete Earth rotation (360°) takes about 24 hours (1 day).

The heliocentric model can be used to explain the motions of celestial objects in relation to the Earth and its movements. (See Section D of this Topic for further explanations of celestial motion models.)

Planetary Motions

Unlike the apparent motion of stars, the movement of planets through the star field is not uniform. The difference in the uniform motion of the stars and the non-uniform motion of the planets is explained by the orbital velocities of the planets as they move around the Sun (heliocentric model).

From Earth observation, the angular diameter of the Moon, Sun, and planets appear to change in size in a cyclic manner. This change in **angular diameter** is due to change in the distance of the Moon, Sun, and planets from Earth. The closer they are to the Earth, the larger they appear and the faster they appear to move in their orbits.

Observations of the planets indicate that each planet, revolving around the Sun, is also rotating on its own axis, much the same as the Earth's motions.

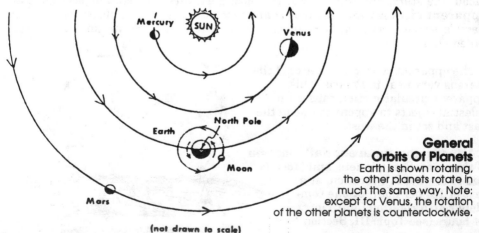

General Orbits Of Planets

Earth is shown rotating, the other planets rotate in much the same way. Note: except for Venus, the rotation of the other planets is counterclockwise.

(not drawn to scale)

Satellite Motions

The Moon is the primary satellite of the Earth, since it revolves around the Earth. (The Earth can be considered a satellite of the Sun for the same reason.) As the Moon revolves around the Earth, it seems to have the same motion characteristics as the other celestial objects. Based on Earth time, one complete revolution of the Moon takes $27\frac{1}{3}$ days (sidereal month).

The angular diameter of the Moon changes as it orbits the Earth. This apparent change in the Moon's diameter is due to the actual orbit of the Moon being slightly elliptical, not circular. (See Section D of this Topic for a further explanation.)

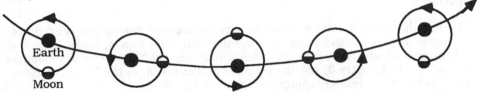

Moon (satellite) Motion Around The Earth

Earth and Moon revolve around their common mass center (**barycenter**).
Note that the Moon does not revolve around the Earth's center.

Phases Of The Moon

As the Moon revolves it passes through a cyclic series of **phases**. The Sun's light rays are always illuminating one half of the Moon. As the Moon revolves around the Earth, the Earth observer sees varying amounts of the illuminated portion of the Moon.

The Moon makes a complete phase cycle around the Earth in $27\frac{1}{3}$ days. However, as the Moon revolves around the Earth, the Earth is moving in its orbit around the Sun, constantly changing the relative positions of the Sun, Earth, and Moon.

Therefore, it takes longer (approximately 2 days) for the Moon to complete its full revolution cycle from new Moon to new Moon. The time for this cycle is $29\frac{1}{2}$ days, known as a synodic month.

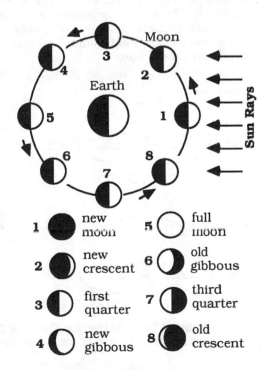

1 new moon
2 new crescent
3 first quarter
4 new gibbous
5 full moon
6 old gibbous
7 third quarter
8 old crescent

Sun Motion

Our Sun appears to move much the same as the other celestial objects, in an arc from east to west, sunrise to sunset. This apparent path varies with the seasons.

The seasons of the Earth are the direct result of the relative position of the Sun in the sky at different times of the year. Because of the Earth's tilt ($23\frac{1}{2}$ ° from a perpendicular, drawn to the plane of the orbit), the Sun's rays are only **perpendicular** (directly overhead) at **noon**, between the $23\frac{1}{2}$° North latitude and the $23\frac{1}{2}$° South latitude during the year. This also causes the sunrise and sunset points on the horizon to vary.

During our summer season, when the Sun has its highest arc, the Sun rises north of east and sets north of west, and the daylight hours are the longest. However, during our winter season, the Sun's path is lower, causing shorter daylight hours, and sunrise and sunset occur south of east and south of west, respectively.

The angular solar diameter varies in a cyclic manner during the year, due to the change in the Earth's distance from the Sun. The Sun appears largest at **perihelion**, which occurs during the winter in the Northern Hemisphere (approximately January 3) and the smallest at **aphelion**, in the Northern Hemisphere during the summer (approximately July 4).

Noon Sun

Local noon is when the Sun reaches its maximum altitude directly over the observer's longitude. When the Sun is in the **zenith** position (directly overhead) at the equator (Sun rays are vertical), daylight and night hours are equal on the Earth. This is known as an **equinox** (generally, March 21st and September 23rd). For the Northern Hemisphere, the first day of summer occurs when the Sun's vertical rays are directly over the $23\frac{1}{2}$° North latitude. During the **summer solstice** (usually, June 21st), the Northern Hemisphere has its longest daylight hours. The **winter solstice** occurs when the Sun's rays are vertical at the $23\frac{1}{2}$° South latitude (generally, December 21st), and the Northern Hemisphere has its shortest daylight hours. The Sun's vertical rays are never seen in the continental United States.

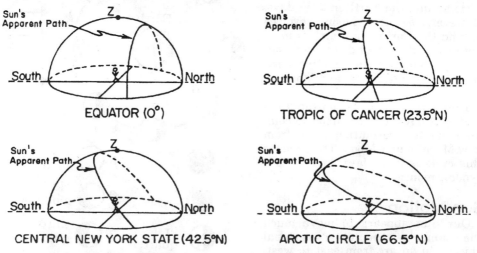

Apparent Path Of The Sun On June 21st, As Seen By 4 Observers
The zenith (Z) is the point in the sky directly over the observer.

B. Terrestrial Observations

What terrestrial evidence suggests Earth motions?

Motion At The Earth's Surface

There are two main motions of the Earth:

- the Earth's **revolution**, a slightly elliptical orbit around the Sun (please see Section D of this Topic), and

- the Earth's **rotation**, a spinning about the north – south axis.

The Earth rotates once in a 24 hour period, turning 360° or 15° per hour, on its axis. The surface rotational velocity is dependent on the observer's latitude. The closer to the equator, the faster the velocity, and the closer to the poles, the slower the velocity. This apparent speed change can be illustrated with the roller skating game "crack the whip," where skaters are strung out in a long line, turning about one skater in the center. The farther the skater is from the center of the spin, the faster the skater must go to maintain the turning chain.

The rotational velocity at the equator is about 465 m/sec. As the observer moves north or south towards the poles, the speed proportionately decreases (at 20° N or S, 437 m/sec, at 60° N or S, 233 m/sec, and essentially 0 m/sec at the poles).

Rotational Evidence

The two main evidences of the Earth's rotation are the apparent motion of the **Foucault pendulum** and the **Coriolis effect**.

1) The **Foucault pendulum** is a freely swinging pendulum, which, when allowed to swing without interference, appears to change direction in a manner that can be predicted.

The apparent direction change is due to the rotation of the Earth below the pendulum. The actual direction of the pendulum does not change.

2) All moving materials on the surface of the Earth (fluids in particular) tend to undergo horizontal deflection, known as the **Coriolis effect**. This predictable effect is caused by the Earth rotating. In the Northern Hemisphere currents and projectiles are deflected to their right, whereas in the Southern Hemisphere, they are deflected to their left. This effect accounts for the motions of winds and ocean currents near the surface of the Earth.

Foucault Pendulum
Back and forth swing
is greatly exaggerated.

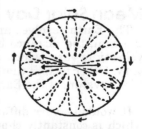

Foucault Pendulum
Pattern of swing is
greatly exaggerated.

Watch the water drain from a sink or shower. If not interfered with, the water will always go down the drain in a clockwise direction (in the Northern Hemisphere). In the Southern Hemisphere, the water will drain counterclockwise.

The traditional model to explain the Coriolis effect is the spinning platform (baseball diamond).

Two persons are standing on a rotating platform (baseball diamond), one at the center and one at the outside edge (perimeter). The platform is rotating counterclockwise. The person at the center throws a ball directly at the other person on the perimeter.

The flight of the ball will be straight, but to the observer the path of the ball will curve behind the person on the edge of the rotating platform. This is because the receiver of the ball has been carried by the moving platform to the right. Had the platform not been moving, the ball would have reached the receiver in a straight line.

The Coriolis effect may also be observed during the liftoff of the Space Shuttle. To the observer, the shuttle appears to be tilted as the Earth rotates below it.

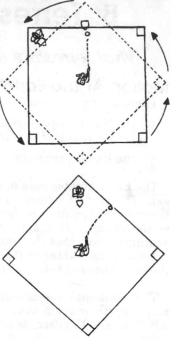

A ball being thrown between two persons riding on a counter-clockwise rotating platform illustrates Coriolis effect.

C. Time

How are frames of reference determined for time?

Time And The Earth's Motions

Frames of reference for time are based upon the motions of the Earth. As previously discussed an **Earth year** is the time required for the Earth to make one complete revolution around the Sun. Although a year is a relatively constant time, a day can be expressed in two ways: the **mean solar day** and the **apparent solar day**.

Mean Solar Day

The Earth revolves around the Sun in an elliptical orbit. As the Earth orbits the Sun, the distance between the Sun and the Earth varies, causing orbital velocity to vary, which in turn causes the time of a day (one Earth rotation) to vary. Other factors that contribute to the variation of time on the Earth include the curvature of the Earth and the change in the Earth's tilt.

It would be very difficult to maintain a clock based on the apparent solar day which is constantly changing. Therefore, the **mean solar day** was established as an average solar day (24 hours). Most days of the year are 24 hours with the rest of the days slightly longer or shorter than 24 hours.

The actual difference in time from one day to the next is measured in fractions of seconds. For example, in the Northern Hemisphere, the longest days are during November through February, averaging up to 15 seconds longer than the shortest days which occur during May through August.

Solar Day

The apparent solar day is the amount of time required for the Earth to rotate from one noon time to the next. It is measured by two successive appearances of the Sun at a given meridian. Because the Earth is revolving as well as rotating, the Earth must actually rotate slightly more than 360° in order to return to noon on successive days.

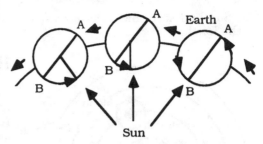

Earth's rotation and revolution is counterclockwise. As the Earth revolves, the Earth must rotate more than 360° for the Sun's zenith to return to the same Earth longitude.

D. Solar System Models

What models explain the observations of celestial and terrestrial motions?

As discussed eariler, there are two models which can be used to explain the apparent motions of celestial objects, the **geocentric** and **heliocentric** models. However, apparent terrestrial motions can only be explained by the heliocentric model.

Geocentric Model

The **geocentric model** (Earth-centered) was an early attempt to illustrate the motions of celestial objects. Although this model **does provide an explanation of the daily motions of the Moon, Sun, and stars**, the geocentric model **does not easily explain the planetary motions**. It also **can not explain** the terrestrial motions, **Coriolis effect**, and the apparent change in the path of a **Foucault pendulum**.

In the geocentric model, the **Earth is a stationary center** with the Sun, Moon, planets, and stars revolving at different velocities in circular orbits. The Moon has the closest orbit to the Earth.

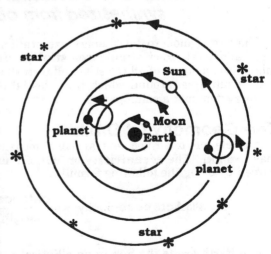

Celestial objects revolving about the Earth as depicted by the Geocentric Model which places the Earth as the center of the Universe.

In the geocentric model, the planets are shown having two orbits. One orbit is around the Earth and a smaller secondary orbit, called an **epicycle**. The epicycles were an attempt to explain the irregular motions (speed and distance changes) of the planets.

The Sun and stars have the largest circular orbits at the greatest distance from the Earth.

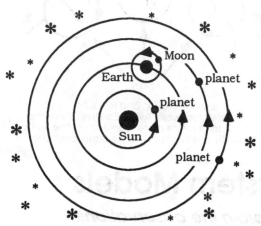

Earth and other celestial objects revolving about the Sun as depicted by the Heliocentric Model which places the Sun as the center of the Universe.

Heliocentric Model

The **heliocentric model**, meaning Sun centered, is a more recent-model to explain celestial motions. It is less complicated than the geocentric model and also explains terrestrial motions.

The Sun, rather than the Earth, is the "stationary" center of this model. The Moon orbits the Earth while the Earth revolves around the Sun in an elliptical orbit. The planets also revolve around the Sun in elliptical orbits.

Outside of this solar system, the stars appear stationary due to great distances from the Earth. (Note. All stars are actually moving.)

What simple celestial model can be synthesized from observations?

The use of modern tools, space exploration, and mathematics has led to the best and most exact **simple celestial model**. This currently used model explains the motions of all celestial and terrestrial objects. There is a center for the entire universe, around which the celestial objects are all in motion, with our solar system just a small part.

The Geometry Of Orbits

A circle has one fixed central point (focus), but an elliptical geometric figure has two foci. The **eccentricity** or "out of roundness" of an ellipse can be determined by using the following formula:

$$\text{elliptical eccentricity} = \frac{\text{distance between foci}}{\text{length of major axis}} = \frac{d}{L}$$

The Earth orbits the Sun in an elliptical path with the Sun at one of the foci. The other planets of our solar system also orbit the Sun in much the same manner with orbits of varying eccentricities.

There are two very important relationships in the simple celestial model. Both involve the time and distance of a planet's orbit.

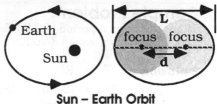

Orbital Speed Of A Planet

Changes in a planet's orbital velocity (time per distance) are based on the following principle:

Sun – Earth Orbit
L = major axis length
d = distance between foci

The areas swept out by an imaginary line connecting the Sun and a planet are equal for equal intervals of time.

In an elliptical orbit, the point at which the planet is closest to the Sun is called its **perihelion**, and the point at which the planet is the farthest from the Sun is its **aphelion**. A planet's fastest orbital velocity occurs at perihelion, and its slowest orbital velocity occurs at aphelion.

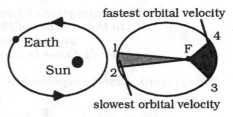

fastest orbital velocity

slowest orbital velocity

Orbital Velocity Comparison
The Earth orbit time between 1-2 and 3-4 is the same. The areas within 1-2-F and 3-4-F are the same.

Period Of A Planet
(Kepler's Harmonic Law Of Planetary Motion)

A planet's **period** is the length of time required for the planet to orbit (revolve) one time around the Sun. Therefore the **period** of a planet **equals one year** for that planet.

The period of any planet is related to the mean radius of its orbit. The further a planet is from the Sun, the longer its period and time to revolve in Earth years. The formula at the right illustrates this relationship.

$$T^2 \propto R^2$$

The square of the planet's period (T^2) is proportional to the cube of its mean distance from the Sun (R^3).

In order to determine this relationship for any planet, the formula at the right can be used.

$$T^2 = R^2$$

Where:
T   is expressed in <u>Earth years</u>, and

R   is expressed in <u>Astronomical Units</u>. (1 Astronomical Unit is the mean distance of the Earth from the Sun – about 150,000,000 kilometers.)

The result should be: 1 = 1

Sample Problem:

Planet "X" has a period of 2.33 Earth years, and a distance from the Sun of 1.76 Earth's average distance.

Solution: $\quad T^2 = R^3$

$\qquad\qquad T^2 = (2.33)^2 \quad = 5.43 \qquad$ relationship verification:

$\qquad\qquad R^3 = (1.76)^3 \quad = 5.45 \qquad 1 = 1 \text{ (rounded off)}$

Force And Energy Transformations

Gravitation is the attractive force between any two objects anywhere. The principle that governs the gravitational attraction in the simple celestial model is:

> *The gravitational force is directly proportional*
> *to the product of the masses of the objects*
> *and inversely proportional*
> *to the distance between their centers squared.*

This proportion can be expressed in the following formula:

$$F \propto \frac{M_1 M_2}{R^2}$$

Where:

$F \quad = \quad$ gravitational force

$M_1 \quad = \quad$ mass of the first object

$M_2 \quad = \quad$ mass of the second object

$R^2 \quad = \quad$ distance between the centers of the two objects

$\qquad$ (Note: d^2 is also used to denote the distance
$\qquad$ - *Earth Science References Tables*)

A **cyclic energy transformation** between kinetic (energy of motion) and potential energy (stored energy) takes place as the Earth orbits the Sun, resulting in a change in the Earth's speed.

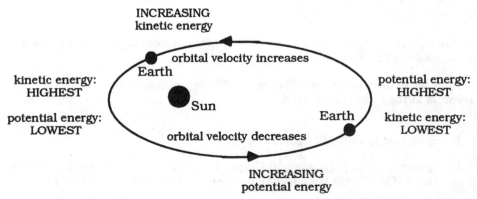

Cyclic Energy Transformation As the Earth orbits away from the Sun . potential energy increases and kinetic energy decreases. The reverse is true as the Earth moves back towards the Sun.

Since the gravitational force acting on the Earth is greatest when the Earth's orbit is closest to the Sun, the orbital velocity of the Earth is increased as it revolves around the Sun. As the Earth speeds up, some of the Earth's **potential energy is converted to kinetic energy**. However, the Earth is not *pulled* into the Sun, since the centrifugal effect on the Earth is also increased, *swinging the Earth around the Sun*.

This centrifugal effect causes the Earth to react as if it had been *thrown out of a slingshot*. As the Earth moves away from the Sun, its orbital speed will decrease as the Sun's gravitational force opposes the Earth's motion (away from the Sun). The gained kinetic energy is reconverted back into potential energy, causing the cyclic energy transformation.

In addition to the cyclic energy transformation, the change in the orbital speed of the Earth causes the length of the day to vary.

Skill Assessments

Base your answers to questions 1 through 5 on your knowledge of Earth Science, the Reference Tables, and the diagram at the right. The diagram represents four planets A,B,C, and D, traveling in elliptical orbits around a star. The center of the star and letter f represent the foci for the orbit of planet A. Points 1 through 4 are locations of the orbit of planet A.

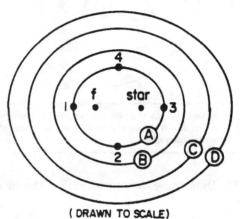

(DRAWN TO SCALE)

1 List the order of the planets from the shortest period of revolution to the longest.

2 If planets A, B, C, and D have the same mass and are located at the positions shown in the diagram, explain in a sentence which planet has the greatest gravitational attraction to the star and why.

3 Write the equation for determining the eccentricity of an ellipse.

4 Using the metric ruler in the Reference Tables and the equation for eccentricity, determine the eccentricity of planet A's orbit.

5 At which numbered position on the orbit of Planet A will Planet A have the greatest orbital velocity? Why?

Base your answers to questions 6 through 9 on your knowledge of Earth Science, the Reference Tables, and the diagrams and table below. Diagram I represents the orbit of an Earth satellite, and diagram II shows how to construct an elliptical orbit using two pins and a loop of string. Table I shows the eccentricities of the orbits of the planets in the solar system.

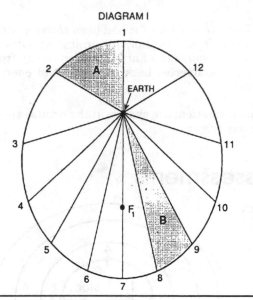

DIAGRAM I

The satellite was at position 1 precisely at midnight on the first day. It arrived at position 2 the next midnight, 3 the next, and so on.

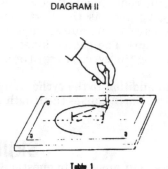

DIAGRAM II

Table 1

Planet	Eccentricity of Orbit
Mercury	0.206
Venus	0.007
Earth	0.017
Mars	0.093
Jupiter	0.048
Saturn	0.056
Uranus	0.047
Neptune	0.008
Pluto	0.250

6   Determine the eccentricity of the satellite's orbit.

7   The Earth satellite takes 24 hours to move between each numbered position on the orbit. In one sentence tell how area A (between positions 1 and 2) compares to area B (between positions 8 and 9)?

8   In one sentence, explain how moving the pins closer together in diagram II would effect the eccentricity of the ellipse being constructed.

9   According to Table I, which planet's orbit would most closely resemble a circle?

Questions For Topic IV

1   A photograph showing circular star trails is evidence that the Earth
    1   rotates on its axis   3   has a nearly circular orbit
    2   revolves around the Sun   4   has a nearly spherical shape

2   How would a three-hour time exposure photograph of stars in the northern
    sky appear if the Earth did *not* rotate?

 (1)   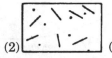 (2)   (3)    (4)

3   The Coriolis effect would be influenced most by a change in the Earth's
    1   rate of rotation   3   angle of tilt
    2   period of revolution   4   average surface temperature

4   The apparent angular diameter of the Sun was
    calculated by an observer in New York State once
    a month for four months. The diameters are
    shown in the data table at the right. Which
    statement is best supported by the data?

Month	Angular Diameter
1	32'16"
2	32'30"
3	32'35"
4	32'31"

    1   The Earth rotates.
    2   The Sun rotates.
    3   The Earth is tilted 23½ degrees.
    4   The distance between the Earth and the Sun varies.

5   Which diagram best represents a heliocentric model of a portion of the solar
    system? [Key: E = Earth, P = Planet, S = Sun. Diagrams are not drawn to
    scale.]

 (1)   (2)   (3)   (4)

6   A Foucault pendulum is set in motion in New York State in a geographic
    north-south direction. Which observation will be made after a period of
    several hours?
    1   The pendulum appears to swing in a wide circle.
    2   The length of the pendulum's swing appears to increase gradually.
    3   The direction of the pendulum's swing appears to change in a predictable
        manner.
    4   The direction of the pendulum's swing appears to change in an
        unpredictable manner.

7   The actual shape of the Earth's orbit around the Sun is best described as
    1   a very eccentric ellipse   3   an oblate spheroid
    2   a slightly eccentric ellipse   4   a perfect circle

8   If viewed from the Earth over a period of several years, the apparent
    diameter of Mars will
    1   decrease constantly   3   remain unchanged
    2   increase constantly   4   vary in a cyclic manner

9   According to the Reference Tables, what is the approximate eccentricity of the ellipse shown at the right?

     (1) 0.50
     (2) 2.0
     (3) 0.25
     (4) 4.0

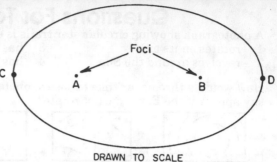

DRAWN TO SCALE

10   The difference between the radius of the Earth and the radius of the Moon is
     (1) 1.74×10^3 km   (3) 6.37×10^3 km
     (2) 4.63×10^3 km   (4) 8.11×10^3 km

11   The phases of the Moon are caused by the
     1   Earth's revolution around the Sun
     2   Moon's revolution around the Earth
     3   Moon's varying distance from the Earth
     4   Sun's varying distance from the Moon

12   The diagram shows the relative position of the Earth, Moon, and Sun for a one-month period.

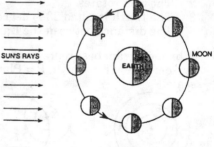

Which diagram best represents the appearance of the Moon at position *P* when viewed from the Earth?

     (1)   (2)   (3)   (4)

13   The diagram shows four different positions (*W, X, Y*, and *Z*) of the Moon in its orbit around the Earth. In which position will the full moon phase be seen from the Earth?

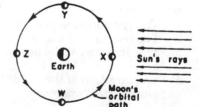

     1   W   3   Y
     2   X   4   Z

14   Planet *A* has a greater mean distance from the Sun than planet *B*. On the basis of this fact, which further comparison can be correctly made between the two planets?
     1   Planet *A* is larger.
     2   Planet *A*'s revolution period is longer.
     3   Planet *A*'s speed of rotation is greater.
     4   Planet *A*'s day is longer.

Questions 15 and 16 refer to the diagram which shows the Earth's orbit and the partial orbit of a comet on the same plane around the Sun.

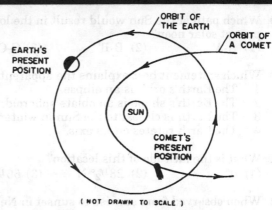

15 Which observation is true for an observer at the Earth's Equator at midnight on a clear night for the positions shown in the diagram?
    1   The comet is directly overhead.
    2   The comet is rising.
    3   The comet is setting.
    4   The comet is *not* visible.

16 Compared with the Earth's orbit, the comet's orbit has
    1   less eccentricity
    2   more eccentricity
    3   the same eccentricity

Questions 17-21 refer to the diagram which represents a plastic hemisphere upon which lines have been drawn to show the apparent paths of the Sun on four days at one location in the Northern Hemisphere. Two paths are dated. The protractor is placed over the north-south line. *X* represents the position of a vertical post.

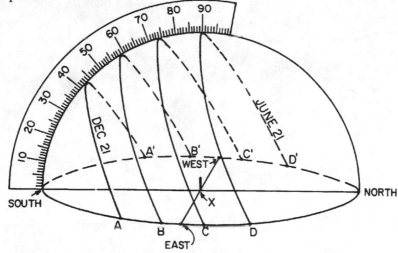

17 For which path is the altitude of the noon Sun 74°?
    (1) A-A'   (2) B-B'   (3) C-C'   (4) D-D'

18 How many degrees does the altitude of the Sun change from December 21 to June 21?
    (1) 43°   (2) 47°   (3) 66½°   (4) 74°

19 Which path of the Sun would result in the longest shadow of the vertical post at solar noon?
(1) A-A'   (2) B-B'   (3) C-C'   (4) D-D'

20 Which statement best explains the apparent daily motion of the Sun?
1   The Earth's orbit is an ellipse.
2   The Earth's shape is an oblate spheroid.
3   The Earth is closest to the Sun in winter.
4   The Earth rotates on its axis.

21 What is the latitude of this location?
(1) 0°   (2) 23½° N   (3) 66½° N   (4) 90° N

22 When observed from sunrise to sunset in New York State, the length of the shadow cast by a vertical pole will
1   decrease, only   3   first decrease, then increase
2   increase, only   4   first increase, then decrease

Energy Budget — Energy In Earth Processes

Vocabulary To Be Understood In Topic V

Absolute Zero
Absorption
Calorie
Conduction, Convection
Conservation of Energy
Convection Cell (current)
Earth Energy
Electromagnet Energy
Electromagnet Spectrum
Friction
Heat Energy
Heat of Fusion, Vaporization
Kinetic, Potential Energy

Latent Heat
Phase Change
Radiation
Radioactivity (decay)
Reflection, Refraction
Scattering
Solar Energy
Solar Electromag. Spectrum
Source and Sink
Specific Heat
Temperature
Transverse Wave
Wavelength

A. Electromagnetic Energy And Energy Transfer

There are many forms of electromagnetic energy, such as heat and light. They all have one thing in common, the capacity (ability) to do work. All of the Earth processes discussed in this book are accompanied by the transfer of energy.

What are the properties of electromagnetic energy?

Electromagnetic Energy

Electromagnetic energy is energy that is radiated in the form of waves. All objects that are not at a temperature of absolute zero radiate electromagnetic energy. Absolute zero is expressed as –273° Celsius or as 0 Kelvin degrees. Theoretically, absolute zero is the point at which all motion of particles and energy transfer stops.

Electromagnetic waves vibrate perpendicular (at right angles) to their direction of travel.

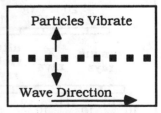

Transverse Wave
At Right Angle
To The Direction

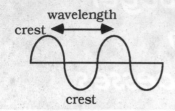

Wavelength Is
Measured Between Crests.

Therefore, they are called **transverse waves**. No mass moves in the electromagnetic waves. Actually, the waves are the result of varying electric and magnetic ("electromagnetic") forces, acting at right angles (perpendicular) to the direction of the source of the waves.

This can be demonstrated using the simple example at the right. Tie one end of a rope to a wall. Take the free end of the rope away from the wall and shake it gently up and down. The rope will move up and down forming transverse waves.

Properties Of Electromagnetic Waves

There are many kinds of electromagnetic energy. Each is distinguished from the others by its **wavelength**, the distance between the crests (peaks) of successive waves.

Waves exist as cycles, meaning they repeat over and over again. The cycles of differing waves are identified by their **frequencies**. When the distance between the wave peaks (wavelength) is compared to time, a frequency results. For example, if an electromagnetic wave repeats 5 times in one second of time, its frequency is 5 or 5 cycles per second. The more often per second a wave repeats, the higher its frequency and the more powerful the electromagnetic energy. The chart (bottom of page) shows a partial listing of different forms of electromagnetic energy, comparing their frequencies. This scale represents the **electromagnetic spectrum**.

Electromagnetic waves are constantly interacting with the environment. When they come in contact with other substances they can be transmitted, refracted, reflected, scattered, and absorbed.

1) **Refraction** (bending or moving in a different direction) occurs when a wave passing through a medium comes in contact with another medium of a different density.

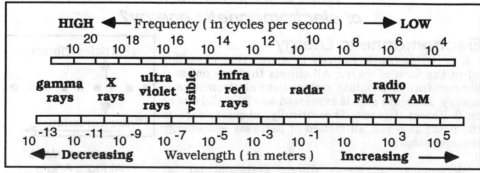

Electromagnetic Spectrum

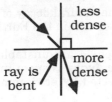

ray is bent

Regular Reflection

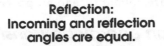

Refraction:
Ray is bent.

Reflection:
Incoming and reflection angles are equal.

Scattering:
Incoming and reflection angles are different.

2) **Reflection** (bouncing off) occurs when a wave contacts a surface and does not pass through. The wave reflects off the surface at the same angle at which it hit the surface.

3) **Scattering** (changing directions – dispersion) occurs when waves reflect in different directions off the surface of a medium.

4) **Absorption** (penetration and capture) occurs when a wave strikes a medium and is taken into that material, not passing through or being reflected.

   A good absorber of electromagnetic energy is a good radiator as well.

5) **Transmission** is when a wave passes through a medium.

The surface of the medium is one factor that determines how much of the electromagnetic energy will be reflected or absorbed. In the case of light energy, if a surface is rough or dark, it is likely to absorb more energy than if the surface is smooth or light. In the south, houses are built with white or light colored shingles on the roofs to reduce the solar heat absorption. In the north, darker shingles are used, allowing more solar heat absorption during the winter months.

Solar Energy

The Sun is the primary source of energy for the Earth's surface. The radiation of the Sun is responsible for driving many of the systems which change the Earth, such as weather and climate patterns.

The **solar electromagnetic spectrum** is similar to the general electromagnetic spectrum, since the Sun radiates almost every kind of wave energy. The principal solar electromagnetic radiations include the visible light spectrum as well as X-rays, ultraviolet rays, and infrared rays.

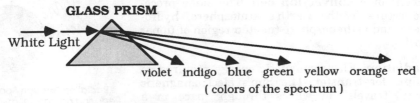

(colors of the spectrum)

Visible Light Spectrum

Earth Energy

The Earth produces some energy itself. For example, the natural decay of radioactive matter is a secondary source of energy for processes within the Earth.

Many atoms, such as uranium-238, radium, and carbon-14 are unstable and break down to form more stable atoms. This **radioactive decay** is spontaneous, occurs at constant rates, is predictable, and is not affected by environmental changes, such as temperature, humidity, and air pressure.

How can energy be transferred?

Energy Transfer

Energy may be transferred in three ways: **conduction**, **convection**, and **radiation**. In matter (solids, liquids, and gases), heat energy may be transferred by either conduction or convection. In space, energy is transferred through electromagnetic wave radiation.

Conduction of thermal (heat) energy occurs as an interaction of matter at the molecular or atomic level. When atoms or molecules collide at an interface, thermal energy is transferred from one atom (or molecule) to another. The efficiency of the energy transfer depends upon the densities of the substances involved. For example, the molecules within a gas are relatively far apart; therefore, the energy transferred is less than within a liquid, where the molecules are relatively close together. Solids have the most compact molecules and have the greatest energy transfer.

Conduction In a Pan
Heat energy is transfered (conducted) through the metal spoon from the hot pan to the hand.

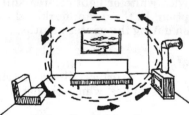

Convection Cell In a Room
Warm air rises, then cools and falls, maintaining the cycle.

Convection is the transfer of thermal energy by the movement of molecules within liquids and gases. The movement is from regions of higher density molecules to lower density molecules. Since warm air is less dense than cool air, the warm air rises displaced by the cool air that settles due to the cool air's greater density.

For example, heated air rises because it is displaced by cooler air. As the cool air comes in contact with the heat source, it is warmed and also rises. The warm air moves away from the heat source, then cools and falls, maintaining the cycle. The resulting circulation is referred to as a **convection current** or a **convection cell**. The same procedure occurs in the Earth's atmosphere, hydrosphere, and asthenosphere (molten region of upper mantle).

Radiation. Electromagnetic energy requires no medium for transfer. The Sun's electromagnetic energy "travels" in waves through space in a straight line until it comes in contact with matter.

Radiation Through Space
Requiring no transfer material, energy is radiated through space.

B. Energy Transformation
What are some energy transformations that can be observed in Earth processes?

Latent Heat

Latent heat is a form of **potential energy** which is absorbed or released when a change of phase occurs. When heat is transferred within the same phase (solid, liquid, or gas), the temperature of the medium is changed.

However, when heat is applied to different states of matter (such as solid to liquid) and there is a **change of phase**, the temperature of the material remains the same, since there is no increase in the kinetic energy of the molecules. Instead, the heat energy is either **absorbed** (**solid to liquid to gas**) or **released** (**gas to liquid to solid**), increasing or decreasing the potential energy of the molecules.

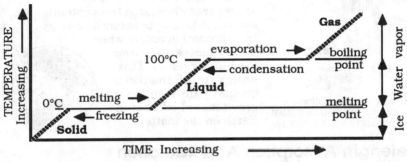

Phase Change Chart For Water

Heat is gained or lost in the phase change. The amount of latent heat is different for various substances and the type of phase change. The amount of heat either gained or lost is equal to the product of the mass of the matter times the latent heat per unit of mass.

Latent Heat Of Water

Melting occurs when water changes states from a solid to a liquid. **Freezing** occurs when water changes from a liquid into a solid. The latent heat for these changes is 80 calories per gram of water and is called the **heat of fusion**. In other words, one gram of ice (solid water) absorbs (gains) 80 calories of heat energy when changed to a liquid. When the phase change is reversed, liquid to ice (**freezing**), 80 calories per gram of heat is released (lost) by the water.

A significantly greater amount of energy is required to change a given mass of water from liquid to vapor than is required to change ice to liquid. **Evaporation**, requiring 540 calories per gram of water (added) (called the **heat of vaporization**) occurs when water in the liquid state is changed to a gas (vapor). When **condensation** (vapor to liquid) occurs, a latent heat of 540 calories per gram of water is released. (Refer to the heating curve for water on next page.)

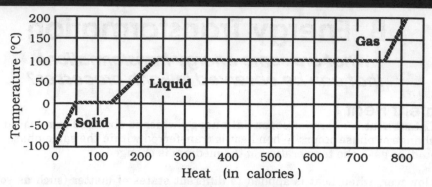

Heating Curve For Water
Shows the heat needed to change 1 gram of water from -100°C (ice) to 200°C (vapor).

Movement Of Matter

The movement of matter toward the Earth's center of mass results in an energy transformation from potential to kinetic. A kinetic to potential energy transformation occurs when matter moves away from the Earth's center. This energy transformation is primarily the result of the gravitational effect of the Earth on the matter.

Energy Transformation
As the rock falls, the result is a transfer of potential to kinetic energy.

Wavelength Absorption And Radiation

As previously discussed, a good absorber of energy is also a good radiator. The characteristics of the surface of the material receiving the electromagnetic energy determine the quantity and type of energy absorbed or reflected.

The Earth's surface often absorbs solar electromagnetic energy (generally strong, short wavelengths) and converts the energy to longer wavelengths (having less strength) which are reradiated. For example, solar ultraviolet rays (short wavelengths) are absorbed by the Earth and reradiated as infrared (long wavelengths). Gases in the Earth's atmosphere absorb much of the infrared energy. This is referred to as the **greenhouse effect**.

Friction

Friction is a resistance to relative motion between surfaces in contact. **Energy transformation occurs at interfaces where there is friction**. For example, there is an interface between a stream bed and the flowing water of the stream. At this interface between the moving water and the sand and rocks of the stream bottom, some of the kinetic energy of the moving stream is lost to the stream bed by friction.

C. Energy Relationships In Earth Processes

What inferences can be drawn about the total energy within a closed system?

Conservation Of Energy

Energy flows from sources to sinks. An **energy source** is a location with a higher concentration of energy than the surrounding area, from which energy will flow towards an **energy sink**, an area of lower energy concentration. For example, the air around a hot wood stove (source) has the highest concentration of heat energy in the room. The heat will flow from around the stove to other parts of the room (sink) that are cooler with lower concentrations of heat energy.

The **principle of the conservation of energy** states that energy can neither be created nor destroyed – energy can be changed only in form. This principle can be observed in closed system, where, **the amount of energy lost by a source equals the amount of energy gained by a sink**.

The amount of energy needed to produce an equal temperature change in equal masses of different materials varies with the materials. The amount of heat energy needed to raise the temperature of one gram of a material one degree on the Celsius scale is called **specific heat**. The higher the specific heat a material has, the more energy it takes to heat it and the longer the time to cool it.

Liquid water has the highest specific heat capacity among all of the naturally occurring materials on Earth (1.0 calories/gram/1°C). (See the Earth Science Reference Tables for the specific heats of other materials.)

Heat Lost Or Heat Gained

The heat (measured in calories) lost or gained by a material is proportional to the product of the mass and the temperature change in the material times the specific heat. To determine the amount of heat lost or gained, use the following formula:

Heat (cal) = Mass x Temperature Change x Specific Heat

The heat lost or gained in a phase change is equal to the product of the mass times the change in potential energy per unit mass.

Latent Heat { Solid ↔ liquid: Heat (cal) = mass x heat of fusion
Liquid ↔ gas: Heat (cal) = mass x heat of vaporization

Skill Assessments

Base your answers to questions 1 through 5 on your knowledge of Earth Science, the Reference Tables, and the graph which shows the temperatures recorded when a 1 gram sample of water was heated from −100°C to +200°C. The water received the same amount of heat every minute.

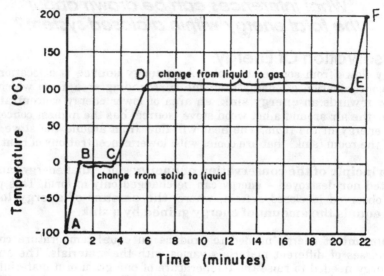

1   How many minutes did it take the water to reach a temperature of 75°C?

2   Between which two consecutive points was the greatest amount of energy being absorbed?

3   In a sentence or two explain why the temperature did not change between points *D* and *E* even though energy was being added.

4   In a sentence or two explain why the temperature changed more rapidly from points *A* to *B* than from points *C* to *D*.

5   How much heat energy would be needed to change the temperature of 10 grams of water from points *C* to *D*?

6   What is the total amount of heat energy needed to change the water from point *A* to point *F*?

Questions For Topic V

1   Which part of the solar electromagnetic spectrum has the maximum intensity?
   1   visible light radiation   3   ultraviolet radiation
   2   infrared radiation   4   X-ray radiation

2   Short waves of electromagnetic energy are absorbed by the Earth's surface during the day. They are later reradiated into space as
   1   visible light rays   3   ultraviolet rays
   2   infrared rays   4   X-rays

The diagram shows part of the electromagnetic spectrum.

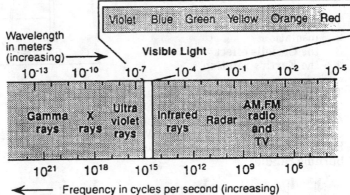

3   Which form of electromagnetic energy shown on the diagram above has the shortest wavelength and the highest frequency?
   1   AM radio   3   red light
   2   infrared rays   4   gamma rays

4   An object that is a good absorber of electromagnetic energy is also a good
   1   reflector of electromagnetic energy
   2   refractor of electromagnetic energy
   3   radiator of electromagnetic energy
   4   convector of electromagnetic energy

5   As the ability of a substance to absorb electromagnetic energy increases, the ability of that substance to radiate electromagnetic energy will
   1   decrease
   2   increase
   3   remain the same

6   Which type of surface would most likely be the best reflector of electromagnetic energy?
   1   dark-colored and rough   3   light-colored and rough
   2   dark-colored and smooth   4   light-colored and smooth

7   At which temperature will iron radiated the *least* electromagnetic energy?
   (1) 0°C   (3) 0°F
   (2) 230 K   (4) 32°F

8   Assuming masses are equal, which substance will require the most heat energy in order to be heated from 40°C to 50°C? [Refer to the Reference Tables.]

    1   basalt   3   lead
    2   iron   4   water

9   The diagram at the right shows a container of water that is being heated.

    The movement of water shown by the arrows is most likely caused by

    1   density differences
    2   insolation
    3   the Coriolis effect
    4   the Earth's rotation

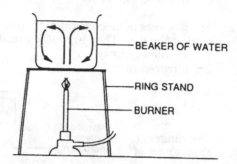

10   The diagram represents a laboratory model used to demonstrate convection currents. Each model shows a burning candle in a closed box with two open tubes at the top of the box. Which diagram below correctly shows the air flow caused by the burning candle?

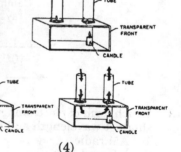

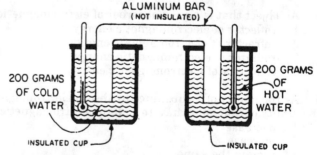

    (1)   (2)   (3)   (4)

Base your answers to questions 11 and 12 on the diagram that illustrates equipment used to perform an *open-system* heat transfer investigation under ordinary classroom conditions.

11   If the initial temperature of the cold water was 10°C and the initial temperature of the hot water was 86°C, what were the most likely temperature readings after 12 minutes?

    (1)   22°C for the cold water and 74°C for the hot water
    (2)   16°C for the cold water and 74°C for the hot water
    (3)   6°C for the cold water and 82°C for the hot water
    (4)   4°C for the cold water and 92°C for the hot water

12 The greatest amount of heat energy transferred between the hot and cold
    water is transferred by the process of
    1   conduction   3   absorption
    2   convection   4   radiation

13 Each arrow in the diagram represents a process involving a phase change of
    water.

    Each process can take place only if
    1   mass and volume remain constant
    2   heat energy is added or released
    3   the number of molecules is increased or decreased
    4   the dew point temperature is increased or decreased

14 The change from vapor phase to liquid phase is called
    1   evaporation   3   precipitation
    2   condensation   4   transpiration

15 Which process results in a release of latent heat energy?
    1   melting of ice   3   condensation of water vapor
    2   heating of liquid water   4   evaporation of water

16 How many calories of latent heat would have to be absorbed by 100 grams of
    liquid water at 100°C in order to change all of the liquid water into water
    vapor at 100°C?
    (1) 100 calories   (3) 1,000 calories
    (2) 8,000 calories   (4) 54,000 calories

17 As heat energy is added to an open container of boiling water the
    temperature of the boiling water will
    1   decrease   2   increase   3   remain the same

18 Two identical towels are hanging on a clothesline in a sunny location. One
    towel is wet. The other is dry. What is one reason that the wet towel feels
    cooler than the dry towel?
    1   Water in the wet towel is evaporating.
    2   Water in the wet towel prevents absorption of heat energy.
    3   The dry towel receives more heat energy from the Sun than the wet
         towel does.
    4   The dry towel has more room for heat storage than the wet towel does.

19 In a closed system, the amount of energy lost by an energy source
    1   is less than the amount of energy gained by an energy sink
    2   is greater than the amount of energy gained by an energy sink
    3   equals the amount of energy gained by an energy sink

20 According to the Reference Tables, how many calories of heat energy must be added to 5 grams of liquid water to change its temperature from 10°C to 30°C?
(1) 5 calories
(2) 100 calories
(3) 20 calories
(4) 150 calories

Base your answers to questions 21 through 25 on your knowledge of Earth Science, the Reference Tables and the diagram and table.

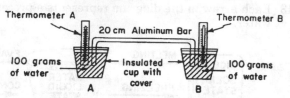

A student used the apparatus shown in the diagram above to perform an experiment. At the beginning of the experiment, the temperature of the water was 90°C in cup *A* and 10°C in cup *B*. The student took readings from the two thermometers for 14 minutes and recorded the following information.

Time (Minutes)	Cup A Temp (°C)	Cup B Temp (°C)
0	90	10
2	87	10
4	84	10
6	81	11
8	78	13
10	75	15
12	72	17
14	69	19

21 Which graph best represents the relationship between time and the temperature of the water in cup *A*?

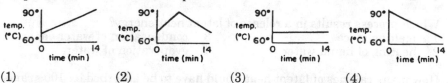

(1)   (2)   (3)   (4)

22 Which statement best explains why the amount of heat energy gained by cup *B* is less than the amount of heat energy lost by cup *A*?
1 Water heats more slowly than it cools.
2 The aluminum bar is losing some heat to the air.
3 The thermometer in cut *A* generated heat energy.
4 Heat energy flows from heat sinks to heat sources.

23 What was the amount of heat gained by the 100 grams of water in cup *B* during the 14 minutes?
1 90 calories
2 110 calories
3 900 calories
4 1,100 calories

24 The difference between the amount of heat energy lost by cup *A* and the amount of heat energy gained by cup *B* could be decreased by
1 replacing the 20-cm aluminum bar with a 10-cm bar
2 using more water in cup *A*
3 using metal cups instead of insulated cups
4 lowering the room temperature to 10°C

25 If all of the heat lost by cup *A* were gained by cup *B*, what would be the highest possible temperature of the water in cup *B*?
1 19°C   2 40°C   3 50°C   4 89°C

Insolation And The Earth's Surface

Vocabulary To Be Understood In Topic VI

Absorption of Solar Energy
Aerosols
Angle of Insolation
Direct (Vertical) Rays
Duration, Intensity of Insolation
Greenhouse Effect

Incident Insolation
Perpendicular Insolation
Random Reflection
Radiative Balance
Seasons
Reflection and Scattering

A. Earth's Surface Insolation

The primary source of energy for the Earth is **solar radiation**. Solar radiation has its greatest intensity occurring in the visible wavelength (about 50%) of the solar electromagnetic spectrum. The other forms of insolation include ultraviolet, infrared, X-rays, and other longer and shorter wavelengths.

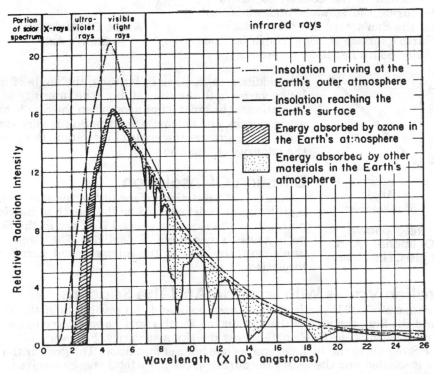

The term **Insolation** comes from **In**coming **Sol**ar Radi**ation**. It is that portion of the Sun's radiation which is received at the Earth's surface.

What are some factors which affect insolation?

Insolation

Angle of Insolation. The intensity of insolation in any area of the Earth's surface increases as the angle of insolation approaches perpendicular. At an angle of 90° to the Earth's surface, the solar radiation is concentrated in a relatively small area of the Earth. Therefore, where the Sun's rays are vertical, the maximum amount of solar energy is received at the surface.

The angle of insolation changes on any particular date with an increase or decrease in latitude. For example (in the diagram at the right), at 23½° North, the Sun's rays are vertical and therefore have maximum intensity. The intensity of insolation per unit area will decrease with increasing latitude towards the North and South Poles, because the insolation received is spread over a greater area.

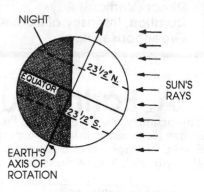

At the equinoxes, the Sun's rays are perpendicular to the equator (0° latitude), while during the summer and winter solstices, the Sun's rays are perpendicular to the Earth's surface at 23½° North Latitude and 23½° South Latitude, respectively.

**Angle Of Insolation
To Earth's Axis**

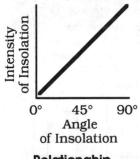

**Relationship
Of Intensity And
Angle Of Insolation**

In addition to changes in latitude, the angle of insolation varies with the time of day. The intensity of the insolation is the greatest at noon when the Sun is highest in the sky and is least when the Sun is very low in the sky. The greater the angle of insolation, the greater the intensity of insolation.

Duration Of Insolation

Duration of insolation refers to the number of daylight hours. It is determined by the length of the Sun's path across the sky. For every 15° of arc in the Sun's path, there is one hour of daylight. The longer the Sun's path, the greater the intensity and duration of insolation.

Duration Varies With Latitude And The Seasons

Because of the 23½° tilt of the Earth's axis and the revolution of the Earth around the Sun, the rays of the Sun hit the Earth's surface with varying angles, and the length of the Sun's path across the sky also varies. For every 15° of arc in the Sun's path, the surface receives one hour of insolation. The greater the angle of insolation and the longer the duration, the more total energy received.

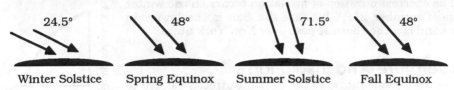

Angle Of Insolation Related To Seasons
(Arrows indicate the Sun's rays at noon for 42° North latitude.)

The seasons are the result of yearly cyclic changes in the duration and intensity of solar radiations or **insolation**. These cyclic changes result from:

1) **Inclination (tilt) of the Earth's axis**, at $23\frac{1}{2}°$, allows the vertical rays of the Sun to fall on different latitudes of the Earth (between the Tropics of Cancer — $23\frac{1}{2}°$ North and Capricorn — $23\frac{1}{2}°$ South).

2) **Parallelism of the Earth's axis**. Since the Earth's axis is always pointing into space in the same direction, the axis of the Earth at any given point in the Earth's orbit around the Sun remains parallel to the axis at any other given point of the orbit.

3) **Revolution of the Earth** causes the Sun's perpendicular rays to fall on different Earth latitudes between $23\frac{1}{2}°$ North and South Latitude.

4) **Rotation of the Earth** causes the alternation of day and night.

In the northern mid-latitudes (42° North), the **maximum insolation** occurs about June 21st (summer solstice). The angle of insolation at noon is $71\frac{1}{2}°$, and the duration of insolation is 15 hours.

Average insolation occurs on or about March 21st and September 23rd (equinoxes). The angle of insolation at noon is 48°, and the duration of insolation is 12 hours.

The **least insolation** occurs about December 21st (winter solstice). The angle of insolation at noon is $24\frac{1}{2}°$, and the duration of insolation is 9 hours.

Summer And Winter Solstice

In the Northern Hemisphere, the Sun's path is the longest on the summer solstice (June 21st). Therefore, New York State has the greatest duration of insola-

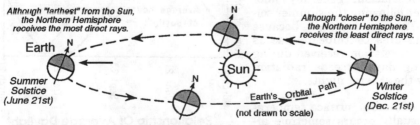

Summer and Winter Solstice Showing Relative Earth – Sun Positions

tion. The shortest duration of insolation occurs on the winter solstice (December 21st), because the Sun is the lowest in the sky and has the shortest path over New York State.

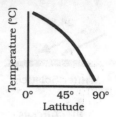

Relationship Of Average Annual Surface Temp. And Latitude

Temperature And Insolation

The surface temperature of the Earth is directly related to insolation received at the surface.

Incoming solar radiation raises the Earth's surface temperature (heat gain). However, terrestrial radiation causes a cooling effect on the Earth's surface (heat loss).

Surface temperature is a result of the relationship between heat gained and lost. Where insolation exceeds radiation, the temperature rises. If more heat is lost (radiation) than is gained (insolation), the temperature at the Earth's surface decreases.

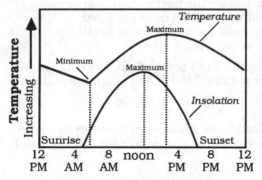

Relationship Of Average Daylight Hours And Daily Temperatures

Daily Temperatures

During the day the **maximum surface temperature** usually occurs sometime **after maximum insolation**, usually during the early afternoon. The minimum surface temperature usually occurs approximately an hour before sunrise, due to the continuous loss of heat during nighttime (radiation). This is called **temperature lag**.

There are other factors which affect the maximum and minimum temperatures, such as cloud cover. Clouds tend to reduce the amount of heat lost due to surface radiation, but also reduce the amount of insolation reaching the Earth's surface.

Yearly Temperatures

During the year the maximum surface temperature most often occurs sometime after the maximum insolation, generally midsummer following the summer solstice. This is the case, because temperatures continue to rise as long as insolation received during the long days exceeds radiation during the shorter nights.

The minimum surface temperature usually occurs sometime after the minimum angle of insola-

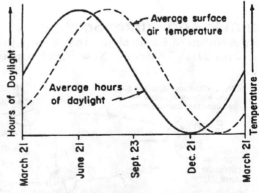

Relationship Of Average Daylight Hours And Annual Temperatures

tion, soon after the winter solstice. This is called **seasonal lag**. Again there are factors (such as weather systems) which affect the yearly maximum and minimum temperatures. As long as radiation during the long nights exceeds insolation received during the shorter days, temperature will continue to decrease (for further discussion see Topic VII.).

Absorption

Since the atmosphere is largely transparent to visible radiation, most of the Sun's visible (light) radiation reaches the Earth's surface. But, because the atmosphere selectively absorbs solar radiations, most of the Sun's ultraviolet radiation is absorbed by the atmosphere's ozone, and much of the infrared radiation is absorbed by the atmosphere's carbon dioxide and water vapor.

To a large extent the surface of the Earth itself tends to control temperature changes through absorption and radiation. Water surfaces heat more slowly than land surfaces, and water also tends to hold heat longer. Therefore, land surface temperatures change more rapidly and change to a greater degree than water surface temperatures. The resulting temperature differences between land and water masses effect greatly average Earth temperatures.

Reflection

Particles in the atmosphere and materials on the surface of the Earth greatly affect the amount of insolation that reaches the Earth's surface and is absorbed. For example, clouds may reflect approximately 25 percent of the incident insolation.

The reflectivity of the Earth also depends upon both the surface and the angle of insolation. The greater the angle of insolation, the greater the absorption, but as the angle of insolation decreases, the reflection becomes greater. A surface of ice and/or snow may reflect almost all of the **incident insolation**.

Scattering

Aerosols (finely dispersed particles, such as water droplets and dust) in the atmosphere cause a random reflection, or **scattering**, of insolation. The amount of insolation reaching the Earth's surface decreases as the amount of random reflection increases.

For example, a volcanic eruption may put sufficient ash into the air to decrease the amount of insolation reaching the Earth's surface, resulting in a temperature decrease.

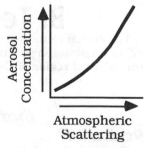

**Scattering Effect
Of Aerosols In
The Atmosphere**

Energy Conversion

Not all of the insolation is directly reflected or radiated as heat energy. Instead, some of the insolation is converted into potential energy by the evaporation of water and the melting of ice. This energy conversion does not change the surface temperature.

Greenhouse Effect

As previously discussed, much of the radiant energy from the Sun 1) passes through the Earth's atmosphere, 2) is absorbed by the atmosphere or Earth's surface, or 3) is reradiated by the Earth's surface back into the atmosphere. Visible light is the most intense energy (from the Sun) to be absorbed by the Earth's surface. Most of the infrared radiation that is reradiated back into the Earth's atmosphere is absorbed by water vapor and carbon dioxide. This process which causes the atmosphere to be heated is known as the **greenhouse effect**. The warmed atmosphere acts as a "thermal blanket," reducing the loss of energy to space and raising the temperature of the Earth's surface.

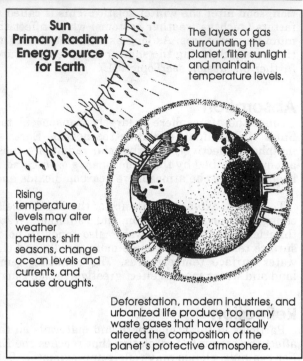

Sun Primary Radiant Energy Source for Earth

The layers of gas surrounding the planet, filter sunlight and maintain temperature levels.

Rising temperature levels may alter weather patterns, shift seasons, change ocean levels and currents, and cause droughts.

Deforestation, modern industries, and urbanized life produce too many waste gases that have radically altered the composition of the planet's protective atmosphere.

The Greenhouse Effect On Earth

The **greenhouse effect** is the result of the conversion of shortwave energy (insolation, such as visible radiation) to longer wave radiation, infrared, which is prevented from escaping the Earth's atmosphere.

B. Terrestrial Radiation

Terrestrial radiation refers to the electromagnetic energy that the Earth gives off to its atmosphere and space. This radiation from the surface is nearly all in the infrared region of the electromagnetic spectrum.

What are some factors that affect terrestrial radiation?

Radiation

The maximum intensity of outgoing radiation from the Earth's surface, **terrestrial radiation**, is in the infrared region of the electromagnetic spectrum. As previously discussed, some of this infrared, heat, energy is absorbed by the atmosphere, and the balance of the energy escapes into space. The **gases** (i.e., water vapor and carbon dioxide) are the primary absorbers of infrared energy in the Earth's atmosphere. They are responsible for the greenhouse effect. Since carbon dioxide is a good absorber of infrared energy, scientists today are concerned about the long term effects of increasing amounts of carbon dioxide in the atmosphere.

Radiative Balance

When the average Earth temperature remains stable, the Earth is said to be in **radiative balance**, gaining as much energy as it gives off. In general, there are only two periods of time in the year when the Earth is in radiative balance, during the midsummer and the midwinter, sometime after the summer and winter solstices when insolation and radiation are in equilibrium.

Comparing the radiative balance of the Earth over long, medium, and short time periods indicates the following:

- **Long-term Measurements** (thousands of years) of worldwide surface temperatures indicate that the Earth *is not* in radiative balance. For example, during the Pleistocene Epoch Age, the estimated average temperatures of the Earth varied approximately 5°C to 10°C cooler, resulting in the Ice Age.

- **Intermediate-term Measurement** (decades) of worldwide surface temperatures indicate that the Earth *is* in radiative balance. Although temperature records show that there are cyclic variations in surface temperatures, there are general trends towards stability of Earth surface temperatures.

- **Annual Measurement** of worldwide surface temperatures indicates that the Earth *is not* in radiative balance. Daily, weekly, monthly, and seasonal temperatures are in a constant state of change, often as a result of weather systems.

Skill Assessments

Base your answers to questions 1 through 4 on your knowledge of Earth Science and the diagram which represents a plastic hemisphere upon which lines have been drawn to show the apparent path of the Sun on four days at a location in New York State. Two of the days are December 21 and June 21. The protractor is placed over the north–south line.

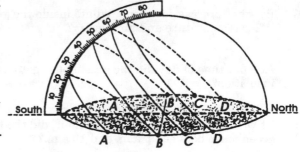

1   What is the solar noon altitude of the Sun for path C–C'?

2   Label the December 21 and June 21 paths on the diagram. How many degrees does the altitude of the Sun change from December 21 to June 21?

3   Which path was recorded on a day that had twelve hours of daylight and twelve hours of darkness? How can you tell?

4   In a sentence or two explain why the apparent path of the Sun changes during the year.

Base your answers to questions 5 through 12 on your knowledge of Earth Science, the Reference Tables and the diagram and data below. The diagram represents a closed glass greenhouse located in New York State. The data table shows the air temperatures inside and outside the greenhouse from 6 a.m. to 6 p.m. on a particular day.

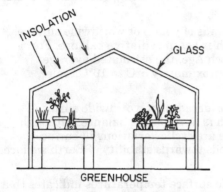

AIR TEMPERATURE

Time	Average Outside Temperature	Average Inside Temperature
6 a.m.	10°C	13°C
8 a.m.	11°C	14°C
10 a.m.	12°C	16°C
12 noon	15°C	20°C
2 p.m.	19°C	25°C
4 p.m.	17°C	24°C
6 p.m.	15°C	23°C

Use one or more complete sentences to answer each of the following questions:

5   Where and when did the highest temperature occur?

6   By what process did the energy come from the Sun to the greenhouse?

7   By what process does air circulate inside the greenhouse due to differences in air temperature and air density?

8   At what time of day did the greenhouse get maximum intensity of insolation?

9   At what rate, in degrees Celsius/hour, did the temperature rise inside the greenhouse between 8 a.m. and 10 a.m.?

10   What happens to the intensity of insolation received by the greenhouse from February 1 to April 1?

11   Explain why the inside of the greenhouse heats up between 6 a.m. and 2 p.m.

12   Explain how the "greenhouse effect" heats the troposphere.

Questions For Topic VI

1   Electromagnetic energy that reaches the Earth from the Sun is called
   1   insolation   3   specific heat
   2   conduction   4   terrestrial radiation

2   The factor that contributes most to the seasonal temperature changes during
   one year in New York State is the changing
   1   speed at which the Earth travels in its orbit around the Sun
   2   angle at which the Sun's rays strike the Earth's surface
   3   distance between the Earth and the Sun
   4   energy given of by the Sun

3   The map shows isolines of average
daily insolation received in calories
per square centimeter per minute at
the Earth's surface. If identical
solar collectors are placed at the
lettered locations, which collector
would receive the *least* insolation?
   (1) *A*   (3) *C*
   (2) *B*   (4) *D*

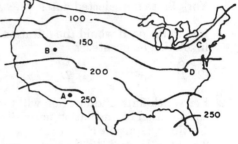

4   The intensity of the insolation that reaches the Earth is affected most by
   1   the angle at which the insolation strikes the Earth's surface
   2   changes in the length of the Earth's rotational period
   3   the temperature of the Earth's surface
   4   the distance between the Earth and the Sun

5   Which graph best illustrates the relationship between the angle of insolation
   and the time of day at a location in New York State?

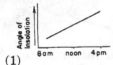 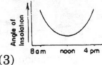 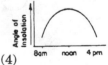

   (1)   (2)   (3)   (4)

6   The tilt of the Earth on its axis is a cause of the Earth's
   (1) uniform daylight hours   (3) 24-hour day
   (2) changing length of day and night (4) 365½-day year

7   Over a period of one year, which location would probably have the greatest
   average intensity of insolation per unit area? [Assume equal atmospheric
   transparency at each location.]
   1   Tropic of Cancer (23½°N)   3   the Arctic Circle (66½°N)
   2   New York City (41°N)   4   the North Pole (90°N)

8   Compared to the polar areas, why are equatorial areas of equal size heated
   much more intensely by the Sun?
   1   The Sun's rays are more nearly perpendicular at the Equator than at the
      poles.
   2   The equatorial areas contain more water than the polar areas do.
   3   More hours of daylight occur at the Equator than at the poles.
   4   The equatorial areas are nearer to the Sun than the polar areas are.

9   What happens to the angle of insolation between solar noon and 6 p.m. in New York State?
1   It decreases steadily.   3   It remains the same.
2   It increases steadily.   4   It first increases and then decreases.

10   On which date does the maximum duration of insolation occur in the Northern Hemisphere?
1   March 21   3   September 23
2   June 21   4   December 21

11   The diagram represents a model of the Sun's apparent path across the sky in New York State for selected dates.

For which path would the duration of insolation be greatest?
(1) *A*   (3) *C*
(2) *B*   (4) *D*

12   For which date and location will the longest duration of insolation normally occur?
1   June 21, at 60°N.   3   December 21, at 60°N.
2   June 21, at 23½°N.   4   December 21, AT 23½°N.

13   The diagram represents four positions of the Earth as it revolves around the Sun.

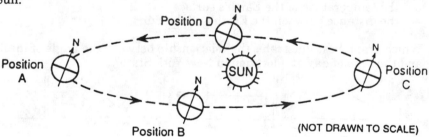

At which position is the Earth located on December 21?
(1) *A*   (2) *B*   (3) *C*   (4) *D*

14   In New York State, it is observed that the north-facing slopes of mountains usually retain their snow later in the spring than the south-facing slopes. This is caused by the fact that the north slopes of the mountains
1   are protected from the prevailing south winds
2   receive greater rainfall
3   usually are steeper
4   receive less insolation than the south slopes

15   A greenhouse stays relatively warm on a sunny winter day. Which statement best explains this fact?
1   The greenhouse traps long-wave infrared radiation.
2   Sunlight is changed to shorter wavelength radiation.
3   The plants growing in the greenhouse produce heat.
4   Glass is an excellent absorber of sunlight.

16 In New York State, when do the highest air temperatures for the year
   usually occur?
   1   a few weeks before maximum insolation is received
   2   a few weeks after maximum insolation is received
   3   at the time that maximum insolation is received

17 Which form of radiation given off by the Earth causes heating of the Earth's
   atmosphere?
   1   infrared   3   visible
   2   ultraviolet   4   X-ray

18 Which gases in the atmosphere best absorb infrared radiation?
   1   hydrogen and nitrogen   3   water vapor and carbon dioxide
   2   hydrogen and carbon dioxide   4   water vapor and nitrogen

19 Which model best represents how a greenhouse
   remains warm as a result of insolation from
   the Sun?

KEY:
~~~~ Short waves
—— Long waves

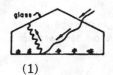
(1)

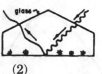
(2)

(3)

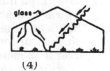
(4)

20 Why are some scientists concerned about an increase in the amount of
   carbon dioxide in the atmosphere?
   1   An increase would change the decay rate of radiocarbon.
   2   An increase would cause more ultraviolet energy to strike the Earth's
        outer atmosphere.
   3   An increase would cause energy to flow from energy sinks to energy
        sources.
   4   An increase could cause an
        overall heating up of the Earth's
        atmosphere.

21 Two identical glass containers were
   placed in direct sunlight. The first
   container was filled with air and
   the second container was filled with
   a mixture of air and additional
   carbon dioxide. Each container was
   sealed with a thermometer inside.
   Temperatures were recorded at
   2-minute intervals, as shown in the
   data table.

	Temperature (°C)	
Time (minutes)	Container 1 (Air)	Container 2 (Air + CO_2)
0	24°	24°
2	25°	26°
4	26°	29°
6	27°	32°
8	28°	33°
10	29°	35°

   Which statement best explains the
   results of this activity?
   1   Carbon dioxide is a good absorber of infrared radiation.
   2   Carbon dioxide causes a random reflection of energy.
   3   Carbon dioxide has no effect on the atmosphere's energy balance.
   4   Carbon dioxide converts some energy into potential energy.

22 The addition of dust to the atmosphere by volcanic eruptions will most likely
    increase the amount of
    1   radiant energy reflected by the atmosphere
    2   insolation absorbed by the Earth's surface
    3   solar radiation absorbed by the oceans
    4   ultraviolet rays striking the Earth's surface

23 What is the primary reason New York State is warmer in July than in
    February?
    1   The Earth is traveling faster in its orbit in February.
    2   The altitude of the noon Sun is greater in February.
    3   The insolation in New York is greater in July.
    4   The Earth is closer to the Sun in July.

24 If an object with a constant source of energy is in radiative balance, the
    temperature of the object will
    1   decrease   2   increase   3   remains the same

25 The graph indicates the
    average number of daylight
    hours and the average
    surface air temperature
    over a 12-month period at a
    specific location on the
    Earth.

    Based on the graph, the
    highest average surface air
    temperature occurs

    1   on June 21
    2   between June 21 and
         September 21
    3   on December 21
    4   between December 21
         and March 21

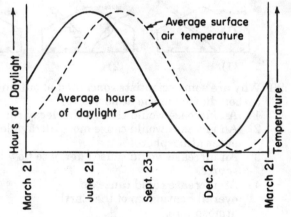

26 The diagram at the
    right shows the path of
    the Sun on March 21
    at a location on the
    Earth.

    According to the
    diagram, which side of
    the house will receive
    the most direct
    insolation?

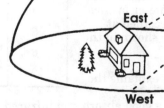

    1   north   2   east   3   south   4   west

Energy Exchanges In The Atmosphere

Vocabulary To Be Understood In Topic VII

Absolute Humidity
Adiabatic Changes
Air Mass
Atmospheric Variables
Cloud
Cold, Warm, Occluded Front
Condensation
Condensation Nuclei, Surface
Convection Cell and Current
Convergence, Divergence
Cyclone, Anticyclone
Dew Point Temperature
Evaporation, Transpiration
Evapotranspiration
Frictional Drag
HIGH and LOW Pressure
Isobar and Isotherm

Moisture
Moisture Capacity
Planetary Wind Belts
Precipitation
Pressure Gradient
Probability of Occurrence
Relative Humidity
Saturation Point
Saturation Vapor Pressure
Source Region
Storm Track
Sublimation
Synoptic Analysis
Vapor Pressure
Water Vapor
Weather
Wind

A. Atmospheric Variables

The changing factors on the Earth that cause weather conditions are called **atmospheric variables**, and include temperature, air pressure, moisture, air movements, and atmospheric transparency.

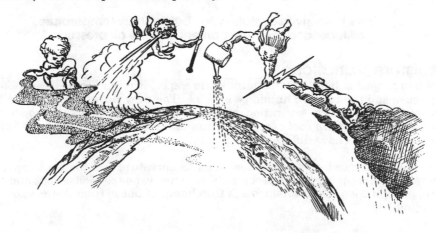

Weather is the local short term condition produced by the interrelationships of these variables in the atmosphere. *Most weather occurs in the troposphere* (the atmospheric layer closest to the Earth's surface) since the tropopause is the limit of convection currents.

Variations in the intensity and duration of incoming solar radiation (**insolation**) control much of the Earth surface conditions through variations in heating and cooling. In this way, insolation either directly or indirectly influences the atmospheric variables. Since these variables are related in a complex way, accurate weather prediction is also complex. The relationships between these atmospheric variables are often expressed as the probability of occurrence. (**Weather prediction**: the short term forecasting of temperature, air pressure, moisture and air movement changes).

What are some relationships between atmospheric variables that can be observed locally?

Temperature Variations

The temperature at the Earth's surface is greatly affected by the amount (intensity and duration) of insolation. Other factors, such as different surface materials (land and water) and altitude, also result in a variety of temperature conditions.

Pressure Variations

Air pressure is caused by the weight of the air. Variations in air pressure may be due to temperature, altitude, or moisture content. Since warm air is less dense than cool air, warm air has a lower mass per unit volume (weight). Being less dense, warm air exerts less pressure, and is therefore lighter.

Air pressure also decreases with an increase in altitude because there are fewer gas molecules per unit volume, as you go up in the atmosphere. Moist air also exerts less pressure than dry air. This is due to light water vapor molecules replacing the heavier gas molecules.

There is an inverse relationship between air temperature, altitude, and moisture content and the air pressure.

Moisture Variations

When the air is holding maximum water vapor, it is said to be saturated with moisture and the relative humidity is 100%. The **dew point** temperature is the temperature at which saturation occurs. *Warm air has a greater capacity for holding moisture than cool air*, since warm air is less dense. Therefore, differing air temperatures have different dew points.

As the difference between the dew point temperature and the air temperature decreases, there is a greater probability of water vapor condensation, and therefore, precipitation. **Condensation** is the change of phase from water vapor (gas) to liquid.

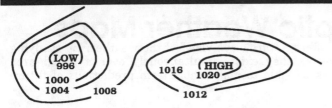

**Isobars
The Relationship
Of LOW And HIGH
Pressure Air Masses
(in millibars)**

Air Pressure Measured

As air is heated and warms, it expands, causing its density and pressure to decrease. The reverse is true of decreasing (cooling) temperatures. Therefore, a change in temperature brings about a change in pressure. Note: Since temperature changes produce pressure variations, then pressure changes also produce temperature variations.

Air pressure is measured by a **barometer** in inches of mercury or millibars. Standard air pressure, also known as one atmosphere, at sea level is 1013.2 millibars (29.92 inches of mercury) of pressure. **Isobars**, lines of equal pressure, are used on a weather map to indicate air pressure regions. On U.S. Weather Bureau maps, the interval between isobars is 4 mb.

Air Movement

One major effect of changes in air pressure is the movement of air from more dense (high pressure) to less dense (low pressure) regions. The horizontal movement of air is called **wind**. Wind speeds are directly related to pressure field gradients. That is, the greater the difference between the air pressures of two areas, the steeper the gradient and the higher the wind speed. Steep pressure gradients are shown on weather maps using isobars that are closely spaced.

Atmospheric Transparency

If the atmosphere consisted only of clear, colorless gases (oxygen, carbon dioxide), almost all of the Sun's rays would reach the surface of the Earth with very few rays scattered, reflected, or absorbed, and the atmosphere would be **transparent**. However, there are aerosols (e.g., dust, water droplets) suspended in the atmosphere which do scatter and absorb radiation.

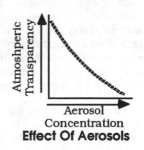

Effect Of Aerosols

Atmospheric transparency varies inversely with the amount of input of materials produced by natural processes and the activities of humans. When Mount St. Helens erupted, tons of volcanic dust were added to the troposphere, decreasing the atmospheric transparency. Pollution from factories, automobiles and homes (for example, smog)

adds more particles to the atmosphere. Variations in the amount and type of atmospheric aerosols may directly affect weather.

When the atmospheric transparency is decreased, usually there is an increase in cloud formation and precipitation. These natural processes tend to cause the atmosphere to periodically clean itself.

B. Synoptic Weather Maps

A synoptic weather map offers a "bird's eye view" of the weather. By studying the atmospheric conditions as they exist simultaneously over a broad area (synoptic observations), it is possible to make short term predictions of future weather conditions.

What air mass characteristics can be determined from synoptic observations?

Air Mass Characteristics

In the atmosphere, an **air mass** is a large body of air having uniform characteristics, including temperature and moisture, that are fairly uniform at any given level. These air masses are identified on the basis of their **average air pressure**, **moisture content**, **winds**, and **temperature**.

Air Mass Source Regions

Air masses have definite characteristics which depend upon the geographic region of origin, called the **source region**. In general, if the origin of an air mass is at *high latitude, it is cold.* If the source region is at *low latitude, it is warm.* For example, in the Northern Hemisphere, cool air masses come mainly from the north, and warm air masses come mainly from the south. An air mass formed over *water is moist,* but if the air mass forms over *land, it is dry.* As an air mass remains stationary in its source region, it becomes more intense and generally larger.

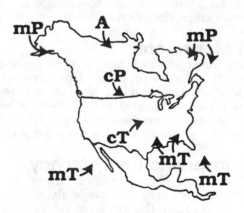

North American Air Masses

On a weather map, air masses are identified and described according to their source regions. An **arctic** air mass is identified with an **A**, a **polar** with a **P**, and a **tropical** with a **T**. In addition, if the source is over *water, it is called maritime* (**m**). If it forms over *land, it is called continental* (**c**).

Identifying an air mass is easy. If the source of an air mass is a northern land area, such as in Canada, it is identified as polar continental (**cP**), and is described as cold and dry. An air mass formed over the Gulf of Mexico would be identified as an **mT** air mass, being warm and moist.

Cyclones And Anticyclones

Cyclone (**Low Pressure System**). In the Northern Hemisphere, air circulates into a low pressure or **cyclone** in a counterclockwise direction. This pattern of air motion in a cyclone may be hundreds of kilometers in diameter. Converging air rises near the center of the low. Cyclones are associated with clouds and precipitation.

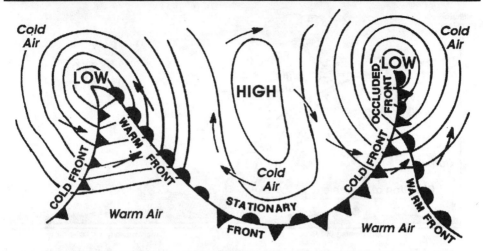

Wind Direction Relationship Between LOW And HIGH Pressure Areas
In the Northern Hemisphere, air currents move around a HIGH in a clockwise direction and around a LOW in a counterclockwise direction. On a Weather Map, similar symbols are used to represent various fronts.

Anticyclone (High Pressure System). In the Northern Hemisphere, the circulation of air in a high pressure air mass or **anticyclone** is clockwise, spiraling outwards. The descending air in anticyclones is associated with cool and clear but often windy weather.

Fronts
A front is the boundary or interface between two air masses on the ground. In the air this boundary is called the **frontal surface**. A front forms between two air masses having differing temperatures. Here along the frontal surface, atmospheric conditions are usually unstable, and precipitation is most probable.

Northern Hemisphere Cyclonic Weather
A **warm air front** forms when the warm air meets and slowly glides up and over the back of cold air. A **cold front** is the result of cold air pushing into a region of warm air. As the fronts are forming, the entire cyclonic storm (low pressure from which the warm and cold fronts extend) generally moves toward the northeast directed by the strong winds high in the troposphere.

Both fronts are associated with clouds and precipitation, because warm air is moving upward, resulting in cooling and condensation. However, at warm fronts stratiform clouds produce steady precipitation, usually of longer duration than cold fronts. The cumuliform clouds associated with the faster moving cold fronts, usually result in brief but heavy precipitation, often accompanied by gusty winds and thunderstorms.

When a cyclone is first developing, it moves rapidly, but as it becomes older, the cyclone slows. Over time, the wedge of warm air tends to narrow, because

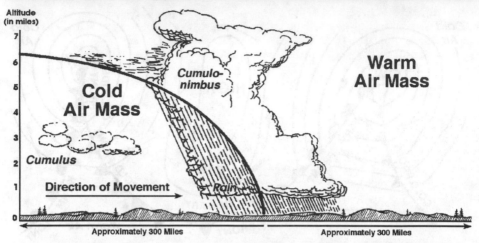

Cold Front showing cold air pushing under warm air and displacing it.

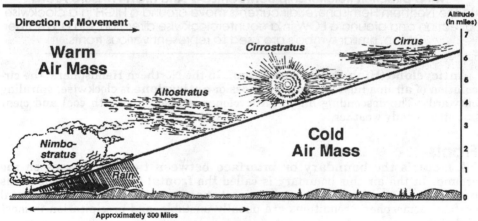

Warm Front showing warm air rising over cold air and displacing it.

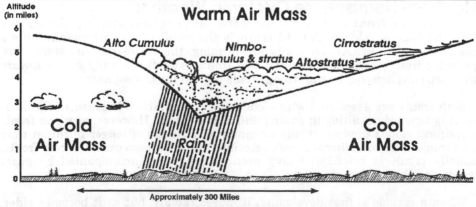

Occluded Front (cold front type) with interface between cool and warm air.

the cold air moves faster than the warm air, forcing the less dense air up. The cold air front overtakes the warm air front ,resulting in an **occlusion**.

An **occluded front** is the interface between a cool and a cold air mass. When the trailing cold front overtakes the warm air front, the two cooler air masses converge and force all of the warm air to rise. Occluded fronts produce some type of heavy precipitation.

A **stationary front** is produced when two air masses of differing characteristics are not moving. The weather of these fronts most closely resembles warm front weather.

Air Mass Tracks

The path that an air mass or storm travels is called the **track**. The rate of movement of a storm can usually be determined and is usually predictable in the continental United States. Tropical air masses generally move northeast, and polar air masses usually move southeast.

In the northern middle latitudes, the prevailing westerly winds blow from west to east and tend to direct the weather patterns. High in the troposphere, there are strong winds capable of slowing down a jet airplane as much as 300 kilometers per hour. These wavelike currents with high winds, called the **jet stream**, move toward the east.

Based on the past observations of weather systems, most storm tracks (paths) can be accurately predicted. But, storm intensity and possible damage from severe storms, such as hurricanes are much more difficult to predict. A potentially dangerous hurricane, such as Andrew (1992), presented a very difficult problem for meteorologists.

The National Hurricane Tracking Service accurately forecasted Andrew's path and strength, but could not predict the exact points of landfall with absolute certainty. When a hurricane tracks close to a coast line, movement of the storm's eye a few kilometers in one direction or another can be the difference between widespread devastation and minor damage.

C. Atmospheric Energy Exchanges

The atmosphere is in a constant state of cyclic changes and exchanges.

How does the atmosphere acquire moisture and energy?

The Input Of Moisture And Energy

Atmospheric moisture content is constantly changing due to evaporation and transpiration. **Evaporation** is the conversion of liquid water to vapor from the Earth's surface, such as soils, lakes, and streams, but primarily from the oceans. Moisture enters the air by means of evaporation and transpiration.

Transpiration is the loss of water from plants through their leaves. Plants absorb water from the ground through their roots, carry it up through their stems and release the water into the air from their leaves. Combined, these two moisture input processes are known as **evapotranspiration**.

Both evaporation and transpiration require energy and constitute an energy input to the atmosphere in the form of more energetic water molecules. The reason for this is that evaporation requires large amounts of energy, approximately 540 calories/gram, to convert liquid water into water vapor (see Topic V).

Vapor Pressure

The rate of evaporation is primarily dependent upon:

- **the amount of energy** available in the location,
- **the surface area** of the body of water, and
- **the vapor pressure** (moisture content) of the air.

The pressure exerted by the water vapor within the atmosphere at a given location is called **vapor pressure**. Near the surface of water (the interface with the air) vapor pressure is the greatest, decreasing with altitude above the water's surface.

When there is very little water vapor in the air, evapotranspiration can occur rapidly, but as the amount of vapor pressure increases, the rate of evaporation decreases until the atmosphere is saturated and cannot hold any additional moisture.

Saturation Vapor Pressure

When the atmosphere contains the maximum amount of moisture that it can hold in the form of vapor, the **saturation vapor pressure** has been reached. This state of the atmosphere is called **dynamic equilibrium**, a balance between the opposing processes of evaporation and condensation.

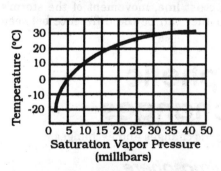

Relationship Of Temperature And Saturation Vapor Pressure

Air temperature has a great affect on the saturation vapor pressure. Warm, less dense air has more space between the molecules of the air, allowing for a greater capacity to hold water. As air is cooled, it becomes more dense, reducing the space between molecules of air. This "squeezes" water vapor out of the air, condensation occurs, and may form precipitation.

Absolute And Relative Humidity

The weight of water vapor contained in a unit volume of air is called **absolute humidity**. Vapor pressure and absolute humidity are directly proportional to each other. **Relative humidity** is the percent of saturation of the air. It is the expression of the ratio between the actual amount of water vapor in the atmosphere and air's capacity at that air temperature.

Relative humidity is determined by using an instrument called a psychrometer. The psychrometer has two thermometers, a dry bulb thermometer and a wet bulb thermometer with a wet wick around the bulb. When whirled in the air, the wet bulb temperature usually drops due to the evaporation of the water, cooling the bulb of the thermometer. The amount of evaporation depends on the moisture content of the air. The lower the moisture content of the air, the more evaporation will occur from the wet bulb, and the lower the wet bulb temperature will be. The bigger the difference between the wet and dry bulb temperatures, the drier the air.

By using the dry bulb temperature and the difference between the wet bulb and dry bulb temperatures (as found on the *Dewpoint Temperatures Chart* in the Reference Tables) the dew point may be determined. The same data may be applied to the *Relative Humidity Chart* to determine the relative humidity. An alternative mathematical method to determine the percent of relative humidity follows:

$$\text{Relative Humidity (\%)} = \frac{\textbf{Vapor Pressure of Dew Point}}{\textbf{Vapor Pressure of Dry Bulb}} \times 100$$

Energy Input

The atmosphere acquires most of its energy (energy input) by **radiation** and **conduction** from the Earth's surface, as well as radiation from the Sun, **insolation**. The rate of insolation and the radiation and conduction from the Earth's surface is not constant. It is related to a number of variables, such as atmospheric moisture and carbon dioxide content. Carbon dioxide and water are good absorbers of radiation from the Earth (see the Greenhouse Effect, Topic VI).

Pollutants, both from natural Earth processes and the activities of humans, increase the atmospheric aerosols and reduce the amount of insolation. Therefore, the **atmospheric transparency** (visibility) is reduced allowing less insolation to reach the Earth's surface. As this occurs, some of the pollutants become the nuclei for condensation, and precipitation. The precipitation, liquid or solid, traps many of the aerosols lowering the atmospheric pollutants and cleaning the air.

Frictional Drag, the friction caused by the Coriolis effect at the interface of the Earth's surface and the atmosphere, is a mechanical means by which the atmosphere acquires energy. The friction produced at this boundary produces some heat which is added to the atmosphere.

Moisture And Energy Transfer

Moisture and energy transfer is accomplished in the atmosphere through three primary means: (1) the density differences, (2) the wind speed and direction, and (3) the adiabatic changes.

How are moisture and energy transferred in the atmosphere?

Density Differences

Density is defined as the ratio of the mass of some object compared to its volume:

Density = mass/volume   or   **D = m/v**

As the moisture content and/or the temperature of air increases, the density of that volume of air decreases. Therefore, density increase is expressed as the inverse (opposite) of temperature and moisture increase (see Topic V).

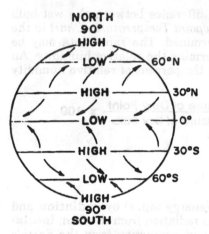

Earth Pressure Belts

One of the results of density differences and the effect of the gravity field is the formation of convection cells. A **Convection Cell** is formed where the air circulates by rising at one place and sinking at another. This same effect can be observed in the ocean and the Earth's mantle, as well as the atmosphere.

Convection cells, or currents, occur because cooler, more dense air sinks towards the Earth's surface, attracted by gravity, causing the warmer, less dense air to rise. Variations in insolation affect convection in the atmosphere.

There are pressure "belts" produced in the atmosphere as a result of convection. Low pressure belts are found at the equator and at the 60° North and South Latitudes, and high pressure belts are found at the 30° North and South Latitudes as well as at the Poles.

In the low pressure belts, air converges and rises; whereas, belts of high pressure are associated with descending air that diverges at the surface. Therefore, regions of low pressure are often called **zones of convergence**, while regions of high pressure are called **zones of divergence**.

Wind Direction And Speed

Movement of air is always from regions of divergence to regions of convergence. As previously discussed, air moves from high pressure to low pressure. As the Earth rotates, wind direction is modified, causing the winds to be deflected to their right in the Northern Hemisphere and to their left in the Southern Hemisphere (Coriolis effect).

The location of the Earth's pressure belts and the effect of the Earth's rotation determine the

		◎	calm
1-2			28-32
3-7			33-37
8-12			38-42
13-17			43-47
18-22			48-52
23-27			53-57

**Wind Speed Symbols
Used On Weather Maps
(in knots 1 flag = 50 knots)**

Planetary Wind and Moisture Belts in the Troposphere

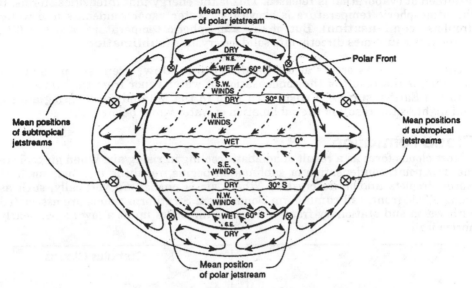

The drawing shows the locations of the belts near the time of an equinox. The locations
shift somewhat with the changing latitude of the Sun's vertical ray. In the Northern
Hemisphere the belts shift northward in summer and southward in winter.

general position and direction of the **planetary wind** circulation (see the illustration above for the position of the planetary wind and pressure belts).

The speed of the winds is directly related to the pressure field gradient. The greater the difference between the high pressure and the low pressure areas, and the less the distance between the pressure centers, the greater the wind speed (velocity).

Along strong fronts, the wind speeds can be very great. However, along weak fronts, the wind speeds are usually much lower.

Adiabatic Changes

The rising (ascending) and falling (descending) air currents cause changes in air temperature by an adiabatic process. An **adiabatic temperature change** is the change in the temperature of air, due to the compression or expansion of the air mass. When air rises, it expands (due to a decrease in pressure), and its temperature decreases. Conversely, as air descends, it compresses (due to an increase in pressure), and its temperature increases.

How are moisture and energy released within the atmosphere?

Condensation And Sublimation

When saturated air is cooled and **condensation surfaces** (*nuclei*) are available, **condensation** will occur. Condensation is the process of water vapor changing phase from a gas to a liquid. When water vapor condenses, the heat energy absorbed at evaporation is released. This is the energy that intensifies storms. If the atmospheric temperature is above 0°C, water vapor condenses into water droplets (**condensation**). But, if the atmospheric temperature is below 0°C, water vapor sublimes directly into solid ice crystals (**sublimation**).

The same process occurs at the Earth's surface. **Dew**, often seen in the early morning, is the result of the condensation of water vapor from the atmosphere onto the Earth's surface. **Frost**, often seen in the fall and early spring after a cold night, is the result of the sublimation of water vapor below 0°C.

Cloud Formation

Most clouds form as a result of adiabatic cooling in rising air. When air cools to the dew point, condensation or sublimation begins, resulting in clouds made of water droplets and/or ice crystals. Where air is pushed up vertically, such as along a cold front, tall cumuliform clouds form. Stratiform clouds are associated with warm and stationary fronts where the air drifts up at a low angle, nearly horizontally.

D. Release Of Moisture And Energy From The Atmosphere

How are moisture and energy released from the atmosphere?

Precipitation

When condensation droplets coalesce (grow together) and become large enough to fall, **precipitation** results. Precipitation may be in the form of liquid water, rain, drizzle, or frozen water, sleet, hail or sublimated water and snow (symbols on next page).

Wind And Water Interaction

The currents on the surface of the ocean are a direct effect of the transfer of energy from the atmosphere to the water and cause the winds that blow over the

Precipitation Symbols

- Rain
- Fog
- Snow
- Hail
- Showers
- Thunder-Storms
- Drizzle
- Sleet

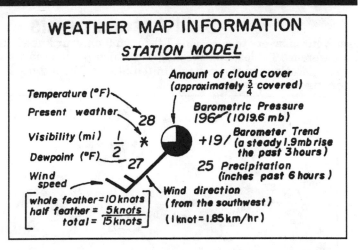

WEATHER MAP INFORMATION

STATION MODEL

Amount of cloud cover (approximately ¾ covered)

Temperature (°F) — 28

Present weather

Visibility (mi) — ½

Dewpoint (°F) — 27

✳

Wind speed

Barometric Pressure 196 (1019.6 mb)

+19/ Barometer Trend (a steady 1.9 mb rise the past 3 hours)

25 Precipitation (inches past 6 hours)

Wind direction (from the southwest)

whole feather = 10 knots
half feather = 5 knots
total = 15 knots

(1 knot = 1.85 km/hr)

water surface. Therefore, a map of surface ocean currents closely resembles the map of planetary winds.

Station Model And Weather Map Reading

The following is a sample station model for a weather map (for further information, see the Weather Map Information section of the Reference Tables).

The **weather station model** describes:
- the sky around the weather observation station, including the amount of cloud cover
- the present weather conditions, including precipitation type and amount
- the visibility (measured in miles)
- the air temperature and dew point in °F
- the wind speed and direction
- the barometric pressure (measured in millibars) and the pressure trend.

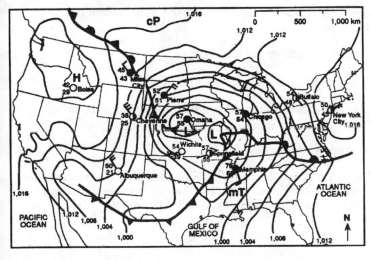

At the left is an example of a Weather Map. The symbols shown on the map identify the many aspects of weather and weather forecasting referred to in Topic VII.

Skill Assessments

Base your answers to questions 1 through 9 on your knowledge of Earth Science, the Reference Tables, and the diagram which represents a section of a weather map for locations in the central United States. The letters *A* through *I* identify reporting weather stations.

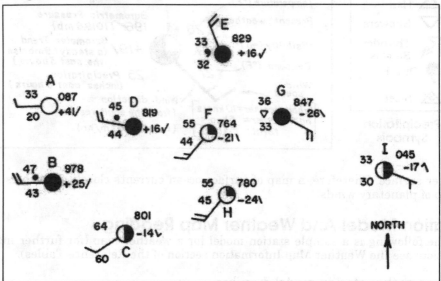

1   On the map draw isolines with an interval of 10°F, beginning with the 40°F isoline.

2   Find the station with the lowest barometric pressure. What is the pressure at that station?

3   Place the letter "L" just to the north of the station with the lowest pressure.

4   Draw and label a cold front and a warm front on the appropriate places extending out of the Low.

5   Which station has the least amount of cloud cover?

6   Which station has a wind from the southeast at 5 knots?

7   Which station shows the pressure has dropped 2.6 mb in the past three hours?

8   What is the air pressure at station *D*?

9   In order to test the rate of evaporation, equal amounts of water are exposed to the open air outside weather stations *B, E, H,* and *I*. In a sentence explain at which station the water will probably evaporate the fastest.

Base your answers to questions 10 through 15 on your knowledge of Earth Science and the satellite photograph of a tropical storm centered in the Gulf of Mexico. An outline of the southeastern United States and the latitude-longitude system have been drawn on the photograph.

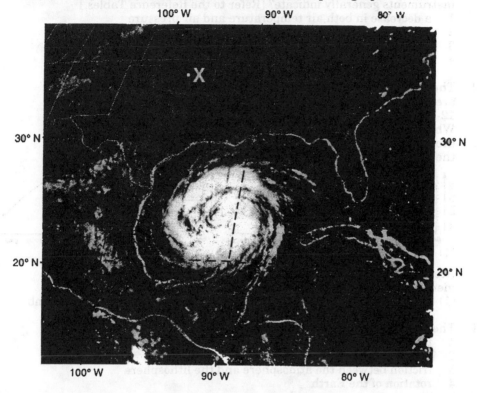

Use one or two sentences to answer the following questions.

10 What is the approximate latitude and longitude of the center or eye of the tropical storm on the satellite photograph?

11 What type of air mass would most likely be associated with the storm ?

12 Describe the weather conditions at point X at the time this photograph was taken.

13 What will happen to barometric pressure along the coast of Texas as the storm approaches?

14 What is the source of energy for this storm?

15 Describe the general direction of movement of the surface winds associated with this tropical storm.

Questions For Topic VII

1   A balloon carrying weather instruments is released at the Earth's surface
    and rises through the troposphere. As the balloon rises, what will the
    instruments generally indicate? [Refer to the Reference Tables.]
    1   a decrease in both air temperature and air pressure
    2   an increase in both air temperature and air pressure
    3   an increase in air temperature and a decrease in air pressure
    4   a decrease in air temperature and an increase in air pressure

2   The graph shows air temperature for an
    area near the Earth's surface during a
    12-hour period.
    Which graph best illustrates the
    probable change in air pressure during
    the same time period?

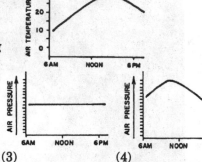

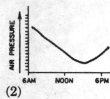

(1)   (2)   (3)   (4)

3   According to the Reference Tables, an air pressure of 29.65 inches of
    mercury is equal to
    (1) 984.0 mb   (2) 999.0 mb   (3) 1001.0 mb   (4) 1004.0 mb

4   The primary cause of winds is the
    1   unequal heating of the Earth's atmosphere
    2   uniform density of the atmosphere
    3   friction between the atmosphere and the lithosphere
    4   rotation of the Earth

5   Wind moves from regions of
    1   high temperature toward regions of low temperature
    2   high pressure toward regions of low pressure
    3   high precipitation toward regions of low precipitation
    4   high humidity toward regions of low humidity

6   The wind speed between two nearby locations is affected most directly by
    differences in the
    1   latitude between the location   3   air pressure between the locations
    2   longitude between the locations   4   Coriolis effect between the locations

7   The Coriolis effect is caused by the
    1   rotation of the Earth on its axis
    2   revolution of the Earth around the Sun
    3   movement of the Earth in relation to the Moon
    4   movement of the Earth in relation to the Milky Way

8   In the Northern Hemisphere, a wind blowing from the north will be
    deflected toward the
    1   northwest   2   northeast   3   southwest   4   southeast

9   The map represents a portion
    of an air-pressure field at the
    Earth's surface. At which
    position is wind speed *lowest*?

    (1) A
    (2) B
    (3) C
    (4) D

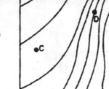

10 Most moisture enter the atmosphere by the processes of
    1   convection and conduction   3   reflection and absorption
    2   condensation and radiation   4   transpiration and evaporation

11 The air temperature and the wet bulb temperature were measured and both
    were found to be 18°C. Two hours later, measurements were taken again
    and the air temperature was 20°C, while the wet bulb temperature
    remained at 18°C. The relative humidity of the air during those two hours
    1   decreased   2   increased   3   remained the same

12 The rate of evaporation from the surface of a lake would be increased by
    1   a decrease in wind velocity
    2   a decrease in the amount of insolation
    3   an increase in the surface area of the lake
    4   an increase in the moisture content of the air

13 As the amount of moisture in the air increases, the atmospheric pressure
    will probably
    1   decrease   2   increase   3   remains the same

14 The two thermometers show the dry-bulb and
    wet-bulb temperatures of the air.

    According to the Reference Tables, what is the
    approximate dewpoint temperature of the air?
    (1) -25°C
    (2) 6°C
    (3) 3°C
    (4) 4°C

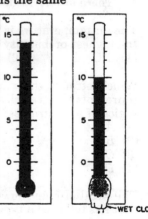

15 Which conditions must exist for condensation to
    occur in the atmosphere?
    1   The air is saturated and a condensation
        surface is available.
    2   The air temperature is above the dewpoint
        and the air pressure is high.
    3   The air is calm and the relative humidity is low.
    4   The relative humidity is low and the air pressure is high.

16 What is the approximate relative humidity if the dry-bulb temperature is
    12°C and the wet-bulb temperature is 7°C?
    (1) 28%   (2) 35%   (3) 48%   (4) 65%

17 Which event will most likely occur in rising air?
1   clearing skies   3   decreasing relative humidity
2   cloud formation   4   increasing temperature

18 Which statement best explains why a cloud is forming as shown in the diagram?
1   Water vapor is condensing.
2   Moisture is evaporating.
3   Cold air rises and compresses.
4   Warm air sinks and expands.

19 On a clear, dry day an air mass has a temperature of 20°C and a dewpoint temperature of 10°C. According to the graph, about how high must this air mass rise before a cloud can form?
(1) 1.6 km   (3) 3.0 km
(2) 2.4 km   (4) 2.8 km

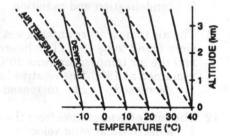

20 Which is a form of precipitation?
1   frost   3   dew
2   snow   4   fog

21 Why is it possible for no rain to be falling from a cloud?
1   The water droplets are too small to fall.
2   The cloud is water vapor.
3   The dew point has not yet been reached in the cloud.
4   There are no condensation nuclei in the cloud.

22 The air temperature is 10°C. Which dewpoint temperature would result in the highest probability of precipitation?
(1) 8°C   (2) 6°C   (3) 0°C   (4) -4°C

23 Which letter on the map at the right represents the area closest to the source region of a cT airmass?
(1) *A*
(2) *B*
(3) *C*
(4) *D*

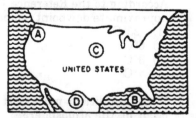

24 The weather map shows a frontal system that has followed a typical storm track

The air mass located over point *X* most likely originated over the

1   northern Atlantic Ocean
2   central part of Canada
3   Gulf of Mexico
4   Pacific Northwest

25 An air mass located over central United States will most likely move toward
    the
    1   northeast   3   northwest
    2   southeast   4   southwest

Base your answers to questions 26 and 27 on the
Reference Tables and the diagram of the station model.

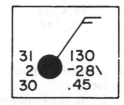

26 The barometric pressure is
    (1) 1013.0 mb   (3) 130.0 mb
    (2) 913.0 mb   (4) 10.28 mb

27 The weather forecast for the next six hours at this station most likely would
    be
    1   overcast, hot, unlimited visibility
    2   overcast, hot, poor visibility
    3   overcast, cold, probable snow
    4   sunny, cold, probable rain

28 Which diagram below best represents the air circulation around a Northern
    Hemisphere low-pressure center?

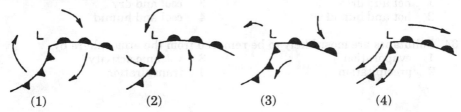

     (1)   (2)   (3)   (4)

29 Cities A, B, C, and D on the weather
    map are being affected by a
    low-pressure system (cyclone)

    Which city would have the most
    unstable atmospheric conditions and
    the greatest chance of precipitation?
    (1) A
    (2) B
    (3) C
    (4) D

30 Which map best represents the normal air circulation around a
    high-pressure air mass located over central New York State?

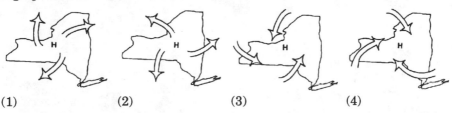

    (1)   (2)   (3)   (4)

31 Which weather station model indicates the greatest probability of precipitation?

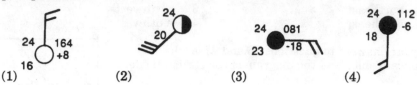

(1)   (2)   (3)   (4)

32 A weather station reporting clear, cold weather with little wind is probably located
1   in the center of a high   3   ahead of a warm front
2   in the center of a low   4   at a cold front

33 An observer reports the following data for a location in New York State:

Air temperature = 35°C
Pressure = 996 mb
Relative humidity = 84%

The weather conditions at this location would best be described as
1   hot and dry   3   cool and dry
2   hot and humid   4   cool and humid

34 Pollutants are most likely to be removed from the atmosphere by
1   evaporation   3   volcanic activity
2   precipitation   4   transpiration

Moisture And Energy Budgets And Environmental Change

Vocabulary To Be Understood In Topic VIII

Actual Evapotranspiration
Aerobic, Anaerobic Bacteria
Arid, Humid Climates
Capillarity
Capillary Water
Change in Soil Storage
Climate
Continental, Marine Climates
Deficit
Ground, Surface Water
infiltration, Permeability
Latitudinal Climate Pattern
Local Water Budget
Orographic Effect

Permeability Rate
Pollution and Pollutants
Porosity
Potential Evapotranspiration
Precipitation
Recharge, Surplus, Usage
Runoff
Soil Storage
Sorted and Unsorted Particles
Stream Discharge
Water Budget, Water Table
Water (Hydrologic) Cycle
Water Purification
Zone of Aeration, Saturation

A. The Earth's Water

As we have learned through the previous Topics, there is a continual movement of water from the atmosphere to the Earth and from the Earth to the
atmosphere. This is known as the **hydrologic** or **water cycle**. The water cycle
includes the phase changes of water and the movements of water above, on, and
below the Earth's surface.

Water Cycle

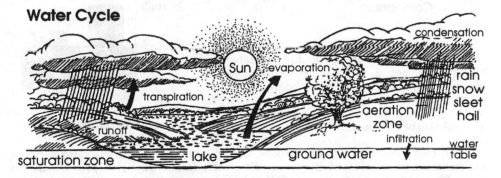

The water cycle is dependent upon the atmosphere to provide water through precipitation to the land and oceans of the Earth.

This precipitation can:

- infiltrate the Earth's surface and become ground water,

- runoff from the surface into streams, lakes, and oceans,

- be stored in the form of ice and snow on the Earth's surface, and/or

- be evapotranspired back into the atmosphere from large bodies of water and plants.

The oceans act as the temporary storage areas for the majority of water moved within the water cycle and act as the major stabilizing factor in the Earth's climates.

Ground Water

How does water move into the Earth?

Porosity

How permeable a material is depends on the **porosity**, which is defined as the percentage of open space between the particles. The porosity of loose material is largely dependent upon the particle's shape, how tightly the material is packed, and the degree to which the particles are sorted.

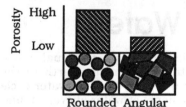

Particle Shape Comparison

For example, water will more easily pass through a cylinder full of round beads, than a cylinder full of square blocks of the same size, since there would be more space between the round beads.

When a test hole is being dug to determine whether or not a septic (waste) field can be placed in a yard, the hole must be dug in "undisturbed" earth, since water will pass through loose soil much faster than the hard packed soil of "undisturbed" ground.

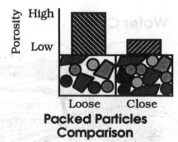

Packed Particles Comparison

When that same septic field is made, the fill used must be gravel and pebbles, rather than smaller sized particles, since the larger particles provide bigger pores for better drainage. The gravel and pebbles will pack less with time, allowing the continuous migration of water. The smaller particles, like clay and silt, will pack and become almost impermeable to water flow.

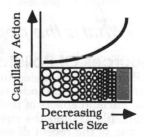

Sorted   Unsorted
Particle Sorting Comparison

Infiltration

The precipitation from the atmosphere can infiltrate (sink into), runoff, or evaporate from the Earth's surface.

Before runoff and evapotranspiration, water will usually infiltrate the Earth's surface. In order for water to move into the surface materials of the Earth, these materials must be permeable and unsaturated. Therefore, loose rock materials, such as soil, sand, and gravel will allow for greater infiltration than the more dense, closely packed particles or the solid rock of the Earth's surface.

Permeability

A material is said to be **permeable** if it allows water to pass though the connecting pore spaces in the material.

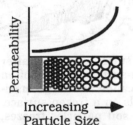

Increasing ➤ Particle Size

Permeability Graph Compared To Particle Size

Water that has infiltrated loose material continues downward until the water reaches the **zone of saturation**. The top of the zone of saturation is called the **water table**. The depth of the water table is dependent upon the type of earth materials, thickness of those materials, the amount of water infiltrating the ground, and the characteristics of the surrounding materials.

For example, the larger the pore spaces between the particles that make up a material, the greater the permeability of that material.

Capillarity

Water can also move upward within a material. This is called **capillarity**. In loose materials, capillarity increases with the decrease in particle size. The finer the loose particle size, the faster and the farther the water can move upward through the material. Along a beach composed of sand, the water will seep up along the shoreline, moving inwards from the water line. However, along a rocky beach, the water will not seep very far inwards from the water line.

Decreasing ➤ Particle Size

Comparison Graph Of Particle Size And Capillary Action

Surface Water

How does water move on the surface of the Earth?

Runoff

When water moves over the surface of the Earth, it is referred to as **runoff**. Surface runoff can occur when rainfall exceeds the permeability rate of the material. For example, a specific sample of soil has a permeability rate of 0.3 liters per hour, but the rate of rain fall is 0.4 liters per hour. The rainfall is greater than the permeability rate by 0.1 liters per hour, which is the amount of runoff from that specific soil sample.

Surface runoff can also occur when the slope of the surface is too great to allow infiltration to occur. Even though the permeability rate of the slope materials may be greater than the rate of rainfall, the water may not have sufficient "standing" time to allow for the infiltration, before it runs off the surface. Water is less likely to soak into a hill side, than to soak into the soil of a flat valley.

If the saturation level of the soil type has reached the surface, such as after long and heavy rains, the added rainfall will run over the surface. This is often the case in severe lowland flooding associated with hurricanes and large maritime tropical storms. Runoff will also occur when rain falls on a frozen surface. The water retained at the points of contact between the soil particles freezes, preventing infiltration of the water which then runs off the surface.

Pollution Of The Earth's Water

Pollution is the contamination of atmospheric, ground, or surface water by the discharge of high levels of harmful or poisonous substances from natural and man controlled processes.

What is the population effect on the Earth's water?

Sources Of Pollutants

Pollutants are added to the hydrosphere through the activities of individuals, communities, and industrial processes. There is heightened public awareness of the harmful effects of pollution on the health, safely, and well being of man, animals, plants, and the environment in general. To a large degree, the awareness is due to the recent discoveries of severely polluted streams, lakes, and ground water all across tall of North America, including the United States, Canada, and Mexico. For example, within the last two decades, increased health problems in the Love Canal area of western New York State led to the discovery of the landfill toxic waste dumps.

For many years industry has had the problem of what to do with the harmful wastes produced during the manufacturing of commercial products. Due to the ignorance of the effects produced by many of these toxic wastes and the economics of dumping these wastes, pollutants have been buried in the ground without the concern or knowledge of ground water systems.

Over the years these poisons seeped into ground water supplies and spread to surrounding areas, contaminating the water sources for communities, even miles away from the actual dump sites. Cleaning up these dumps has proven to be at the least very difficult and in many cases impossible.

A couple of decades ago, the lakes of the Adirondack region were known as some of the best fishing areas in New York State. However, many of the lakes (those over 2000 feet in elevation) now have little or no fish and plant life, due to the pollution of the lakes by **acid rain**.

The wastes from industrial smoke and exhaust emissions (mostly sulfur dioxide and nitrogen oxide) pollute the air to such a great extent that precipitation has become acid. This acid rain falls into the lakes and surrounding water sheds, lowering the pH of the water to such a degree as to make it impossible for animals and plants to survive. Aluminum concentrations toxic to fish are also released by ionized hydrogen exchange, as water percolates through the soil. Attempts to neutralize the lake water have met with only a limited success in small bodies of water. Acid rain may be one of the most serious pollution problems over the next few decades.

Other sources polluting the hydrosphere include, sewage waste from cities, towns, and communities, polluting lakes, streams, and rivers, such as the

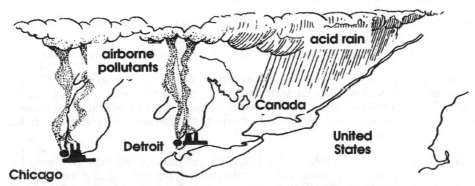

The detrimental effects of acid rain know no boundaries. Airborne pollutants are transported by weather systems across international borders.

Hudson River, heated water waste from nuclear power plants, such as the Shoreham Nuclear Power Plant on Long Island, and the ground water pollution around the industrial plants in the central and southern areas of New York State.

Types Of Pollutants

Hydrospheric pollutants include dissolved and suspended materials such as organic and inorganic wastes, thermal energy effluent from industrial processes, radioactive substances, and the abnormal concentration of various organisms.

- **Organic wastes** include the sewage wastes from communities that contaminate lakes, rivers, and ground water supplies. The biggest danger from organic wastes is the spread of disease in wells and community drinking water supplies.

- **Inorganic wastes** include metal and plastic wastes, pesticide and herbicide often associated with landfills and community dumps, farming and industry. Heavy metals, such as lead, cause poisoning, health problems, cancer and deformities in plants, animals and humans.

- **Thermal wastes** often come from the cooling water of power plants and industrial complexes. The greatest effect of the heated water is in the destruction of aquatic life.

- **Radioactive wastes** come from a variety of sources, such as manufacturing, research, and nuclear power plants. Since many radioactive wastes remain dangerous for very long periods of times (sometimes millions of years), the major problem is how to keep them safe and store the radioactive wastes until they decay to harmless materials. The storage problem has led to other related nuclear material problems, such as safe transportation to storage areas and security in keeping the potentially dangerous materials out of the hands of terrorists.

- **Harmful organisms**, such as disease-causing microörganisms, tend to breed in toxic waste dumps and polluted ground water. Some microörganisms cause cancer, serious diseases, and many health problems.

The release of heated water from industrial and nuclear power plants or the increased activity of aerobic bacteria causes a loss of dissolved oxygen from the water. This form of pollution leads to an increase in the concentration of anaerobic bacteria, a biologic pollutant, which disrupts the normal living and non living cycles in the environment.

If the heated effluent pollutes the breeding grounds of fish, the fish may become extinct or move to other waters, thus destroying fishing and recreational water areas. This form of pollution can be observed in several areas of the Hudson River.

Concentration Of Pollutants

Most often the most severe pollution areas are in the vicinity of population centers. The highest concentrations of pollutants found in the Hudson River are generally located near cities, such as New York City, Poughkeepsie, and Albany. Generally, the same is true around lakes. Ground water pollution tends to be directly proportional to the size of the communities or industrial population. The larger the population, the greater the pollution problem.

Long Range Effects

Left uncontrolled and unchecked, the increase of the hydrospheric pollution could eventually cause the water sources of the Earth to become unfit for human use. Community groups, industry, and government have determined that this should not happen. Beginning with the *Clean Water Act of 1972*, the U.S. has spent billions of dollars to clean up the nation's water supply. In 1980, the United States Congress formed a "Superfund" for the clean up of toxic waste dumps. N.Y. State passed its own *Clean Water Act*, the *Freshwater* and *Saltwater Wetland Acts*. and the environmental quality regulations to protect the waters of New York State and the rights of citizen groups to know the impact on the environment of industrial and public projects.

B. Local Water Budget

A water budget is a model of accounting for moisture income, storage, and outgo for a specific area. The water budget involves many variables, including precipitation, potential and actual evapotranspiration, and moisture storage, usage, deficit, recharge and surplus. The climate of an area (humid or arid) is reflected in its water budget.

Water Budget Variables

How is the water budget influenced by the environment?

Precipitation (P)

The moisture source for the local water budget is **precipitation**. Regardless of what form the precipitation takes (snow, rain, etc.) in the water budget, it is measured in millimeters of water.

Potential Evapotranspiration (E_p)

Potential Evapotranspiration is the maximum amount of water that can be evaporated, if the water is available. The potential evapotranspiration (E_p) of an area is directly proportional to the energy available. Therefore, the greatest potential evapotranspiration occurs in the summer when temperatures are highest.

Actual Evapotranspiration (E_a) is the "real" amount of measured water that is lost from the ground through evaporation and transpiration. The actual evapotranspiration can never exceed the potential evapotranspiration.

Moisture Storage Or Soil Storage (St)

Depending on the characteristics of a soil (porosity, particle size, packing, etc.), a soil under optimum conditions can store a specific maximum quantity of moisture. In computing a water budget, 100 millimeters of water is maximum for storage.

Moisture Usage (negative ΔSt)

The moisture-holding capacity of the soil acts as a reservoir for water stored in the water budget. This stored water is available (if needed) and is taken from the soil storage when the precipitation is less than the potential evapotranspiration ($P < E_p$) during any given month. In New York State a large portion of the yearly total of precipitation falls in the Spring and early Summer. During the drier months of July and August, water is available from the moisture storage in soil. Water evaporating out of ground storage is called **usage**.

Moisture Deficit (D)

A deficit occurs when the potential evapotranspiration is greater than the combined moisture storage and precipitation. A deficit occurs when soil moisture is depleted and there is insufficient precipitation to meet the potential evapotranspiration ($D = St = 0$ and $PE > P$). The amount of deficit equals the difference between E_p and E_a.

Moisture Recharge (positive ΔSt)

The soil moisture is recharged when the precipitation received is greater than the potential evapotranspiration ($P > E_p$), and the ground storage is less than 100 mm. Precipitation is then available to infiltrate and recharge the ground water supply.

Moisture Surplus (S)

There is a surplus of moisture when the soil moisture storage is at its maximum level, and the precipitation is greater than the potential evapotranspiration ($S = St = 100 + P > E_p$). The moisture surplus becomes runoff and contributes to local stream discharge.

Streams

How is the local water budget related to stream discharge?

Stream Discharge And The Water Budget

The rate of streamflow in volume is called **stream discharge** and is measured by determining the amount of water that flows past a specific part of the stream over a specific amount of time. Often the stream discharge is a measurement of the surplus water that drains from the area around the stream. However, if there is no surplus of water in the area of the stream, then stream discharge may be related to the depletion of the ground water from soil storage. This is the case during dry seasons when streams are being fed by the local water table. This is called **base flow**.

Climates And The Local Water Budget

How is the local water budget related to climate?

Climatic Regions

Unlike local weather conditions which are short term effects, a climate is the average weather conditions over much longer periods of time. Local water budget characteristics can be used to distinguish climatic regions (quantitatively, **P/Ep** or **P – Ep**) by determining the **P/Ep** ratio. Humid climates are characterized by the total **P** being significantly greater than the total **Ep**. Climates are arid in regions where the **Ep** is much greater than the **P**.

C. Climate Pattern Factors

Climate is concerned primarily with temperature and moisture conditions. The other variables which affect climate patterns are latitude, altitude, and proximity to large bodies of water, ocean currents, mountain barriers, prevailing winds, and storm tracks.

What factors affect climate patterns?

Latitude

Latitude is the most important factor in determining climate, especially influencing temperature patterns (see Topic VI). Since the duration of insolation at low latitudes is fairly constant, about 12 hours per day, temperature variance is small. Also, at low latitudes, the angle of insolation is always quite high. Therefore, the temperatures remain relatively high. At high latitudes, temperatures vary but remain relatively low due to the generally low angle of insolation.

The duration of insolation varies between 0 hours and 24 hours per day causing a great seasonal variation in temperatures. In general, as the duration of insolation increases, the temperature increases. As latitude increases, average yearly temperature decreases, but the annual temperature range increases.

Elevation

The elevation (altitude) influences the temperature and moisture patterns of a region. The effects of elevation are very similar to those of latitude. Lower elevations are generally more stable in temperature and moisture, and higher elevations have more varied conditions. As the altitude or elevation increases, the average yearly temperature decreases, and the precipitation generally increases.

Large Bodies Of Water

Large bodies of water (large lakes and oceans), ocean currents, and prevailing winds modify the latitudinal climate patterns of their shoreline areas. The slow heating and cooling of large bodies of water cause the land masses near them to have modified temperatures. For example, the **marine climate**, of the northwest coast of the United States, is characterized by cooler summers, warmer winters, than would normally be expected for that latitude. Areas of marine climate have small temperature ranges.

Inland regions are not directly affected by large bodies of water, and have **continental climates**, that are characterized by large yearly temperature ranges with hot summers and cold winters.

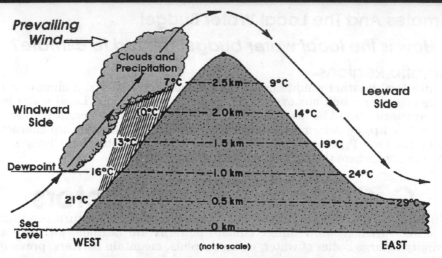

Orographic effect on climatic patterns is caused by "mountain" barriers. Clouds and precipitation form on the windward side of the mountain, whereas, the leeward side is drier and warmer.

Use the diagram above to answer the following questions.

1   What would the approximate temperature of the air be at the top of the mountain?

2   Why do clouds begin to form at the 1.0 km elevation on the windward side of the mountain?

3   The air temperature on the leeward side of the mountain at the 1.5 km level is higher than the temperature at the same level on the windward side. Explain why.

Mountain (Orographic) Barriers

The overall effect of mountains on climatic patterns is called the **orographic effect**. Latitudinal climate patterns are modified by mountains that act as barriers to local weather systems by interrupting the normal path of a prevailing wind. The windward side of a mountain is the side facing the prevailing wind. As the wind hits the windward side of a mountain the air is forced upward and cools adiabatically until the dew point is reached. Then condensation occurs and cooling slows. These conditions cause the water in the air to condense, forming clouds and precipitation, and the rain falls on the windward side of the mountain.

The other side of the mountain, the leeward side, will be drier than the windward side. It is said to be in the *rain shadow*. As the air rises over the mountain and begins to descend on the leeward side, it warms adiabatically (due to the compression), and the air is warmer and drier.

When a mountain is high enough, it can act as a barrier to air masses and prevent warm or cold air from getting past the mountain to the other side. In this case a mountain can be the boundary between two very different climatic areas.

Wind Belts

The planetary winds and pressure belts affect moisture and temperature patterns. If the prevailing winds first cross a large body of water before coming on land, they will bring moisture to the land. If the prevailing winds cross a large land mass, the effect will be more arid. Prevailing winds from tropical areas bring warm air. Winds from polar regions will bring cool air, and winds from the tropic regions will bring warm air.

Storm Tracks

Low pressure systems, cyclones, which effect the temperature and moisture patterns of local areas, seem to follow statistically predictable paths (see Topic VII).

Skill Assessments

Base your answers to questions 1 through 6 on your knowledge of Earth Science, the Reference Tables, and the information and diagrams below which describe an investigation with soils.

Three similar tubes, each containing a specific soil of uniform particle size and shape were used to study the effect that different particle size has on porosity, capillarity, and permeability. A fourth tube containing soil which was a mixture of the same sizes found in the other tubes was also studied. [Assume that the soils were perfectly dry between each part of the investigation.]

Tube	Particle Size (diameter in cm)
A	Fine (0.025 cm)
B	Medium (0.1 cm)
C	Coarse (0.3 cm)
D	Mixed (0.025 to 0.3 cm)

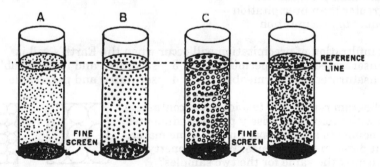

1   The bottom of all the tubes were closed and water was slowly poured into the tubes until it reached the dotted reference line. It took 40 milliliters to fill tube A. How much water did it take to fill tubes B and C? Explain why.

2   In one or two sentences, compare the porosity of tubes C and D.

3   Water was poured into the top of each tube at the same time and allowed to pass through the fine screen at the bottom. Which tube has the greatest permeability?

4   The bottom of tubes *A, B,* and *C* were placed in a shallow pan of water. In which tube did the water rise the highest? Explain why.

5   According to the Reference Tables, what name would be given to the particles the size of those in tube *C*?

6   A handful of material from tube D was dropped into a fifth tube filled with water and allowed to settle. Draw a diagram showing what it would look like. What name is given to this pattern of sorted sediment?

Questions For Topic VIII

1   The diagrams represent two identical containers filled with nonporous uniform particles. The containers represent models of two different sizes of soil particles. Compared to the model containing larger particles, the model containing smaller particles has

     1   less permeability and greater porosity
     2   greater porosity and greater capillarity
     3   less permeability and greater capillarity
     4   greater permeability and greater porosity

2   Dry soil will be recharged with moisture if potential evapotranspiration is
     1   less than precipitation
     2   greater than precipitation
     3   equal to precipitation

3   Most infiltration of precipitation will occur when the Earth's soil is
     1   unsaturated and impermeable   3   saturated and impermeable
     2   unsaturated and permeable   4   saturated and permeable

4   The diagram represents two identical containers filled with samples of loosely packed sediments. The sediments are composed of the same material, but differ in particle size. Which property is most nearly the same for the two samples?

     1   infiltration rate   3   capillarity
     2   porosity   4   water retention

5   Which graph best represents the relationship between air temperature and potential evapotranspiration (E_p) for a given locality?

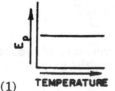

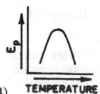

(1)   TEMPERATURE   (2)   TEMPERATURE   (3)   TEMPERATURE   (4)   TEMPERATURE

6   Which diagram below best illustrates the condition of
     the soil below the water table?

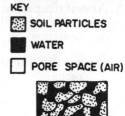

KEY
SOIL PARTICLES
WATER
PORE SPACE (AIR)

 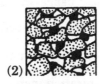 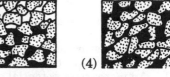

(1)   (2)   (3)   (4)

7   As the amount of precipitation on land increases, the level of the water table
     will probably
     1   fall   2   rise   3   remain the same

8   Which condition exists when precipitation (P) is less than potential
     evapotranspiration (E_p) and soil moisture is zero?
     1   moisture deficit   3   moisture recharge
     2   moisture surplus   4   moisture usage

Base your answers to questions 9 and 10 on the graph of the water budget.

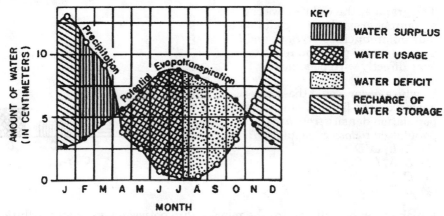

KEY
WATER SURPLUS
WATER USAGE
WATER DEFICIT
RECHARGE OF
WATER STORAGE

9   During which month does the highest average temperature occur?
     1   January   2   April   3   July   4   November

10   During which month is there a water deficit?
     1   February   2   March   3   August   4   December

11   In which area will surface runoff most likely be greatest during a heavy
     rainfall?
     1   sandy desert   3   level grassy field
     2   wooded forest   4   paved city street

12   Which type of soil would water infiltrate most slowly?
     1   silt   3   fine sand
     2   pebbles   4   fine clay

13 As soil temperature decreases from −5°C to −10°C., the infiltration rate of
    surface water will probably
    1   decrease   2   increase   3   remain the same

14 Stream discharge would normally be highest during a period of
    1   recharge   2   deficit   3   usage   4   surplus

15 Why do many streams continue to flow during long periods when there is no
    precipitation?
    1   Soil moisture storage is recharged by vegetation along the stream.
    2   Ground water continues to move into the stream channel.
    3   An increase in potential evapotranspiration increases the stream runoff.
    4   The porosity beneath the stream channels becomes higher than normal.

16 Pollution has become a serious problem chiefly because
    1   modern man's input of pollutants exceeds the rate at which nature can
         remove pollutants
    2   recent changes in landforms have disturbed the rate at which nature can
         remove pollutants
    3   there has been a gradual decrease in the amount of energy available for
         removal of pollutants
    4   there have been recent increases in natural environmental pollutants

17 The graph at the right shows the
    amount of ground water pollution at
    four different cities: A, B, C, and D.
    Ground water pollution tends to
    vary directly with population
    density.
    In which order would these cities
    most likely be listed if ranked by
    population density *from highest*
    population *to lowest* population?
    (1) A, B, C, D
    (2) B, A, D, C
    (3) C, D, A, B
    (4) D, C, B, A

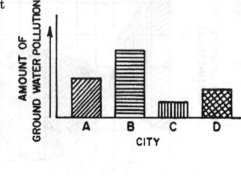

18 Which graph best illustrates the relationship between lake water pollution
    and human population density near the lake?

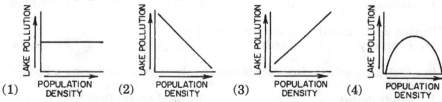

19 The type of climate for a location can be determined by comparing the
    yearly amounts of
    1   precipitation and potential evapotranspiration
    2   soil storage and potential evapotranspiration
    3   precipitation and infiltration
    4   change in soil storage and stream discharge

20 The table shows the precipitation / potential evapotranspiration ratio (P/E_p ratio) for different types of climates.
The total annual precipitation (P) for a city in California is 420 millimeters. The total annual potential evapotranspiration (E_p) is 840 millimeters. What type of climate does this city have?

Climate Type	P/E_p Ratio
Humid	Greater than 1.2
Subhumid	0.8 to 1.2
Semiarid	0.4 to 0.8
Arid	Less than 0.4

1 humid   2 subhumid   3 semiarid   4 arid

21 Which type of climate would most likely be found in an area that has a high potential for evaporation of water but a low actual evaporation of water?
1 polar   2 rain forest   3 desert   4 temperate

22 Which factors have the *least* effect on the climate of a region?
1 latitude and elevation
2 longitude and population density
3 wind belts and storm tracks
4 mountain barriers and nearness to large bodies of water

23 Compared to a coastal location of the same elevation and latitude, an inland location is likely to have
1 warmer summers and cooler winters
2 warmer summers and warmer winters
3 cooler summers and cooler winters
4 cooler summers and warmer winters

24 The diagram represents several locations on the surface of the Earth. Each location is at sea level and is surrounded by ocean water.

The average annual air temperature at point P is most likely higher than the average annual air temperature at point
(1) A
(2) B
(3) C
(4) D

25 The climates of densely populated industrial areas tend to be warmer than similarly located sparsely populated rural areas. From this observation, what can be inferred about the human influence on local climate?
1 Local climates are not affected by increases in population density.
2 The local climate in densely populated areas can be changed by human activities.
3 In densely populated areas, human activities increase the amount of natural pollutants.
4 In sparsely populated areas, human activities have stabilized the rate of energy absorption.

Base your answers to questions 26 through 29 on your knowledge of Earth Science and on the diagram which represents a map of an imaginary continent of the Earth. The continent is surrounded by oceans. Two mountain ranges and five locations, *A* through *E* are shown.

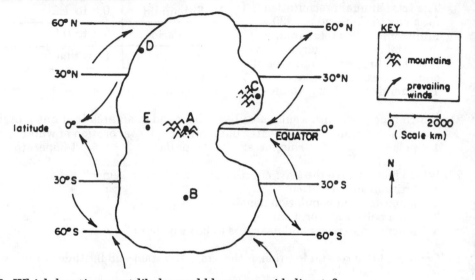

26 Which location most likely would have an arid climate?
(1) *A*   (2) *B*   (3) *C*   (4) *D*

27 As air rises on the windward side of the mountain near location *C*, the air will
1   cool due to expansion   3   warm due to expansion
2   cool due to compression   4   warm due to compression

28 According to this diagram, between which two latitudes are the prevailing southwesterly winds located?
(1) 30°N. and 60°N.   (3) 30°S. and 0°
(2) 30°N. and 0°   (4) 30°S. and 60°S.

29 Which graph best represents the yearly temperature variation for location *B*?

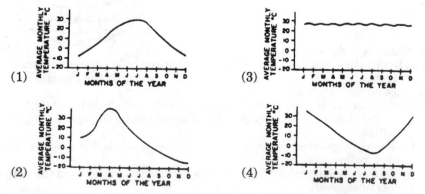

Base your answers to questions 30 through 34 (found on the next page) on your knowledge of Earth Science, the Reference Tables, and the diagrams and map. The graphs in diagram I show the sources of nitrogen and sulfur dioxide emissions in the U.S. Diagram II gives information about the acidity of Adirondack lakes. The map shows regions of the U.S. affected by acid rain.

DIAGRAM I

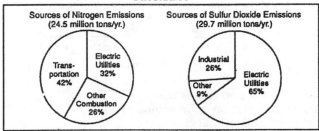

Sources of Nitrogen Emissions (24.5 million tons/yr.)
- Transportation 42%
- Electric Utilities 32%
- Other Combustion 26%

Sources of Sulfur Dioxide Emissions (29.7 million tons/yr.)
- Industrial 26%
- Electric Utilities 65%
- Other 9%

DIAGRAM II

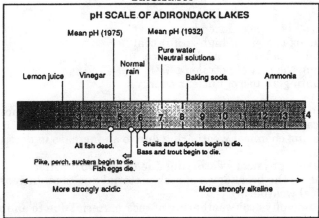

pH SCALE OF ADIRONDACK LAKES

REGIONS OF THE UNITED STATES SENSITIVE TO ACID RAIN

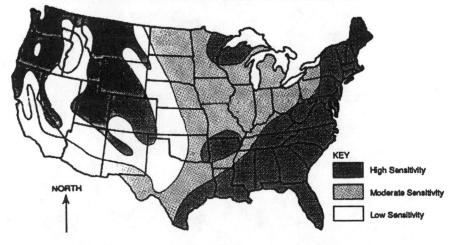

NORTH

KEY
- High Sensitivity
- Moderate Sensitivity
- Low Sensitivity

30 Which pH level of lake water would not support any fish life?
   (1) 7.0   (2) 6.0   (3) 5.0   (4) 4.0

31 Which graph best shows the acidity (pH) of Adirondack lakes since 1930?

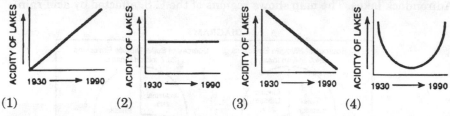

   (1)   (2)   (3)   (4)

32 The primary cause of acid rain is the
   1 weathering and erosion of limestone rocks
   2 decay of plant and animal organisms
   3 burning of fossil fuels by humans
   4 destruction of the ozone layer

33 Acid rain can best be reduced by
   1 increasing the use of high-sulfur coal
   2 controlling pollutants at the source
   3 reducing the cost of petroleum
   4 eliminating all use of nuclear energy

34 In addition to its effects on living organisms, acid rain may cause changes in the landscape by
   1 decreasing chemical weathering due to an increase in destruction of vegetation
   2 decreasing physical weathering due to less frost action
   3 increasing the breakdown of rock material due to an increase in chemical weathering
   4 increasing physical weathering of rock material due to an increase in the circulation of ground water

The Rock Cycle — The Erosional Process

Vocabulary To Be Understood In Topic IX

Chemical, Physical Weathering	Soil Solution
Displaced Sediments	Stream Bed
Erosion	Transported Sediment, Soil
Residual Sediment and Soil	Predominant Agent
Sediment	Transporting Agents, System
Soil Formation, Horizon	Weathering

A. Weathering

The physical and chemical processes that change the characteristics of rocks on the Earth's surface are called **weathering**. In order for weathering to occur, the physical environment of the rocks must change and the rocks must be exposed to the air, water in some form (ice, snow, or liquid), or the acts of man or other living things. Therefore, weathering is the response of rocks to the change in their environment.

What is some evidence that Earth materials weather?

Evidence Of Weathering

The Weathering Processes. When rocks are exposed to the hydrosphere and the atmosphere (water, air, and the substances within them) the physical and/or chemical composition and characteristics of the rocks can change. The end products of weathering are generally called **sediments**, and are classified as boulders, cobbles, pebbles, sand, silt, clay, colloids, and dissolved particles (ionic minerals).

Physical Weathering. In physical weathering, rocks are broken into smaller pieces without changing the chemical nature of the rock. For example, a section of the surface of a mountain may break off because water freezes in a crack in the rock surface, expanding and splitting the rock. The falling rock mass may shatter into boulders and rocks of various sizes. Even though the rock has changed in physical form, it has

Physical Weathering
Action of Freezing on Rocks

not changed chemically and still maintains its original composition. The action of wind or water-carried sediment (erosion) may further abrade (wear away by rubbing) the surface of the rocks forming fine particles.

Chemical Weathering. In chemical weathering, rocks are broken, and the rock material itself is *also* changed. Examples of chemical weathering include:

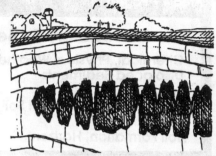

- **oxidation** occurs when oxygen from the air combines with the minerals of the rock to form oxides (for example, iron and oxygen form iron oxide, rust);

- **carbonation** occurs when water containing carbonic acid dissolves the minerals of the rock (for example, the action of carbonic acid on limestone), and;

- **hydration** occurs when minerals, such as mica or feldspar, absorb water weaken and crumble to form clay.

Weathering Agents And Erosional Processes Form Caves

Climate Affects Weathering

Climate affects both physical and chemical weathering. The chart below shows that climate tends to determine the amount, types, and/or rate of weathering. In cold and moist climates, physical weathering is dominant. In warm and humid climates, chemical weathering is dominant. In general, the more moisture available, the more weathering occurs.

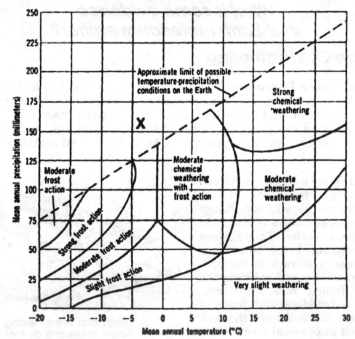

Weathering Rates

The rate at which a material weathers varies inversely with the particle size. In equal amounts of the same kind of rock, the smaller the particle size, the greater the rate of weathering. This is due to a greater surface area in contact with the weathering agents.

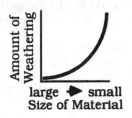

large ➔ small
Size of Material

**Material Size
vs. Weathering**

Rock particles will weather at different rates depending on their mineral composition and its resistance to weathering agents. The harder the mineral, the slower the weathering. The softer the mineral, the faster the weathering.

hardest   softest
Mineral Hardness

**Material
Hardness
vs. Weathering**

Soil Formation

Soil is a combination of particles of rocks, minerals, and organic matter produced through weathering processes. Soil contains the materials necessary to support various plant and animal life.

As a result of the weathering processes and biologic activity, **soil horizons** (layers) develop. These soil horizons vary in depth depending on the amount of weathering, the time over which the weathering occurs, and the climate. Normally, the longer the weathering occurs, the greater the depth of the soil formation.

Soil Formation

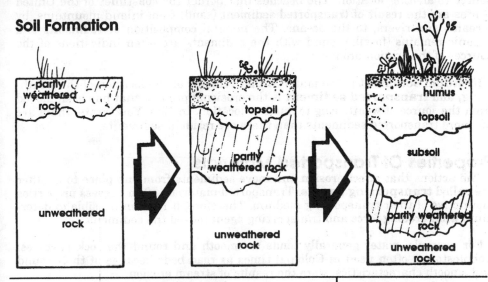

| Immature Soil | Mature Soil |

In addition to weathering, the complex interrelationships of living organisms are significant factors in soil formation. For example, the breakdown and decay of leaves from plants and the life activities of earthworms and other small animals aid in the formation of soil. In addition, the effects caused by frost, rain, and air add to the process of weathering.

Immature soil generally contains partially weathered and unweathered rock. Mature soil has various amounts of topsoil and subsoil containing organic matter, in addition to the weathering products, and partially weathered and unweathered rock. Until a soil has developed a subsoil, it is considered to be immature.

B. Erosion

Erosion refers to the transportation of rock, soil, and mineral particles from one location to another by the action of water, wind, or ice. The motive power behind all of the agents of erosion is **gravity**. It is very important to note that a **transporting system** includes *all of the agents involved in erosion and movement*, including erosion, the transporting agent, energy, and the material moved.

Evidence of Erosion

What evidence suggests that rock materials are transported?

Displaced Sediments

The major evidence of erosion is the displacement of sediments from their source to another location. The beaches that border the coastlines of the United States are the result of transported sediment (sand) from inland mountains, by streams and rivers, to the oceans. The mineral composition of sediments and organic remains (fossils) found with the sediments are often indications of the sources of the erosion products.

Residual sediment is the material that remains at the location of the weathering, and **transported sediment** is the erosional product that has been moved from the source of weathering to another location. In New York State, there are far more transported sediments than there are residual sediments.

Properties Of Transported Materials

The actions that affect erosion and move sediments from one place to another are called **transporting agents**. Transported materials often possess properties distinctive of their transporting medium. Therefore, it is often possible to determine what eroding force and transporting agent moved the sediment.

For example, water generally tends to smooth and round the rock particles. Cobblestones, often used in Colonial times as road beds because of their round and smooth characteristics, were the results of stream erosion.

The longer the water or wind action, the smoother and rounder the sediment becomes. Sediments moved by winds are often pitted, or "frosted."

Glacial products often have surface scratches due to being pushed and scraped by ice. When the action of gravity alone causes erosion, the sediments are sharp edged and angular. This evidence can be observed along highway road cuts and at the base of mountains.

Erosional And Transporting Agents

Factors Affecting Transportation

Gravity, water, wind, ice, and human activities are the main factors affecting the transportation of the sediments of weathering.

How does the transportation of rock materials take place?

Gravity. Most all transporting systems are caused by gravity. Gravity may also act alone, such as when loose pieces of rock on the side of a slope may break away and fall down hill. Gravity also acts with another transporting agent, such as when water transports sediments in a stream.

Water. Running water is the **predominant agent** of erosion on the Earth. There are a number of factors which affect the way in which the running water of a stream transports sediment.

Water Velocity And Sediments

In a stream channel, the average velocity of the running water increases with an increase in the discharge. The **discharge** is the volume of water in the stream at any given location during a specific amount of time. Velocity and discharge of a stream are interdependent. In the spring, the streams usually move faster, due to the greater amount of water volume from melting snow and runoff.

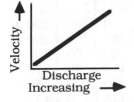
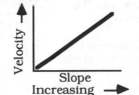
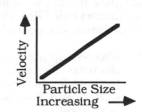

Comparison of stream flow rates to the amount of discharge, the increase in slope angle, and the carried size of particles.

The velocity of the water in a stream is directly proportional to the slope of a stream channel. As the slope of a stream increases, the velocity of the stream increases. For example, the steep sloped mountain streams that feed the Hudson River move faster than the more level Hudson River itself.

As the velocity of the stream water increases, the size of the particles that can be moved by the stream also increases. Streams carry material in solution (dissolved), in suspension, and by rolling or bouncing along the stream bed (**saltation**).

Wind And Ice Erosion

Wind (such as a desert sand storm), ice (such as a glacier), and snow (such as a avalanche) act as transporting agents for rock materials. The factors involved in wind and ice erosion are similar to the factors which affect running water erosion. The steeper the slope of a mountain, the faster a glacier or an avalanche will move and the more and larger the erosional materials that can be carried. Glaciers can transport the largest sized sediments (boulders). Light winds move only the smallest sediments, but strong winds may carry heavier and larger materials, such as sand, but rarely more than a meter or so above the ground (see Topic X for a discussion of material sorting).

Effect Of Humans

Humans add greatly to the natural processes of land erosion through activities, such as highway and industrial construction, destruction of forests from the careless setting of forest fires and uncontrolled tree cutting without reforestation, strip mining, poor landfill projects, and other such activities (see Topic XIV).

Skill Assessments

Base your answers to questions 1 through 6 (below and the next page) on your knowledge of Earth Science and the data table. The table shows the results of an investigation of four different types of rocks, weathering over a time period of 30 minutes. Equal masses of similar-sized samples of rocks A, B, C, and D were placed in identical containers half-filled with water. Each container was shaken uniformly for 5 minutes and the remaining samples of rocks were removed from the water. Their masses were determined and recorded in the data table. The remaining samples of rocks were put back into the containers half-filled with water, and the procedure was repeated five times.

1   Make an appropriate vertical scale for the mass of samples remaining and plot the data for each rock labeling the lines A, B, C, and D respectively (see graph on next page).

DATA TABLE

Mass of Rock Samples Remaining (grams)

Time (min)	A	B	C	D
0	200	200	200	200
5	160	200	120	200
10	125	200	60	195
15	100	190	20	170
20	75	180	0	150
25	55	175	0	135
30	50	175	0	125

2   In a sentence or two, explain the kind of weathering taking place. Which rock sample was the most resistant to weathering?

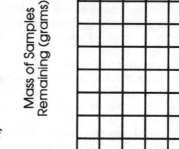

3   How much of the original mass of rock sample *D* was lost after 30 minutes?

4   According to the data table, the mass of the rock samples remaining at the end of 30 minutes was different for each sample. In one sentence, infer what the reason for this difference might be.

5   Suggest what rock sample *B* might be? Why?

6   After 20 minutes, the rate of abrasion decreased for all rock samples. In a sentence or two, explain what the reason might be.

Questions For Topic IX

1   The principal cause of the chemical weathering of rocks on the Earth's surface is
    1   rock abrasion
    2   the heating and cooling of surface rock
    3   mineral reactions with air and water
    4   the expansion of water as it freezes

2   Which type of climate causes the fastest chemical weathering?
    1   cold and dry   3   hot and dry
    2   cold and humid   4   hot and humid

3   Which factor has the least effect on the weathering of a rock?
    1   climatic conditions
    2   composition of the rock
    3   exposure of the rock to the atmosphere
    4   the number of fossils found in the rock

4   The four limestone samples illustrated below have the same composition, mass, and volume. Under the same climatic conditions, which sample will weather fastest?

(1)   (2)   (3)   (4)

5   Water is a major agent of chemical weathering because water
    1   cools the surroundings when it evaporates
    2   dissolves many of the minerals that make up rocks
    3   has a density of about one gram per cubic centimeter
    4   has the highest specific heat of all common earth materials

6   Which property of water makes frost action a common and effective form of
    physical or mechanical weathering?
    1   Water dissolves many earth materials.
    2   Water expands when it freezes.
    3   Water cools the surroundings when it evaporates.
    4   Water loses 80 calories of heat per gram when it freezes.

7   As the humidity of a region decreases, the amount of weathering taking
    place usually
    1   decreases   2   increases   3   remains the same

8   Which erosional force acts alone to produce avalanches and landslides?
    1   gravity   3   running water
    2   winds   4   sea waves

9   According to the Reference Tables, which stream velocity would transport
    cobbles, but would *not* transport boulders?
    (1) 50 cm/sec   (2) 100 cm/sec   (3) 200 cm/sec   (4) 400 cm/sec

10 A river transports material by suspension, rolling, and
    1   solution   2   sublimation   3   evaporation   4   transpiration

11 Which characteristic of a transported rock would be most helpful in
    determining its agent of erosion?
    1   age   3   composition
    2   density   4   physical appearance

12 Which is the best evidence that erosion has occurred?
    1   a soil rich in lime on top of a limestone bedrock
    2   a layer of basalt found on the floor of the ocean
    3   a large number of fossils embedded in limestone
    4   sediments found in a sand bar of a river

13 The map represents a view of a flowing
    stream. The letters identify locations in
    the stream near the interface between
    land and water. At which two locations
    is erosion due to flowing water likely to
    be greatest?

    (1) *A* and *B*
    (2) *B* and *D*
    (3) *A* and *D*
    (4) *B* and *C*

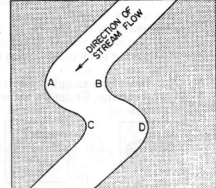

Base your answers to
questions 14 and 15 on
you knowledge of Earth
Science, the Reference
Tables, and the diagram
showing a portion of a
river as seen from above.
The maximum current
velocity of the river is
shown by the dashed line,
XY. Measured rates of
flow are indicated at
points *A*, *B*, *C*, and *D*.

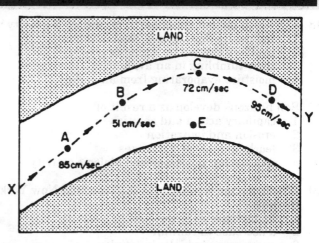

14 Which graph best represents the current velocity of the river at points *A*, *B*,
C, and *D*?

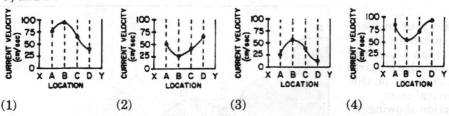

(1)   (2)   (3)   (4)

15 At *X*, particles 0.00001 to 1.0 centimeter in diameter are being carried by the
river. Near which point will 0.1-centimeter particles first be deposited?
1   point *A*   3   point *C*
2   point *B*   4   point *D*

16 Which graph best represents the relationship between stream erosion and
the kinetic energy of a stream?

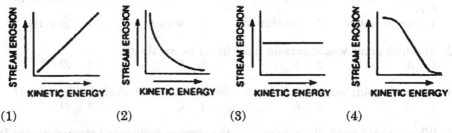

(1)   (2)   (3)   (4)

17 The composition of sediments on the Earth's surface usually is quite
different from the composition of the underlying bedrock. This observation
suggests that most
1   bedrock is formed from sediments
2   bedrock is resistant to weathering
3   sediments are residual
4   sediments are transported

18 For which movement of earth materials is gravity *not* the main force?
1   sediments flowing in a river
2   boulders carried by a glacier
3   snow tumbling in an avalanche
4   moisture evaporating from an ocean

19 Soil horizons develop as a result of
1   capillary action and solution
2   erosion and ionization
3   leaching and color changes
4   weathering processes and biologic activity

20 Which change would cause the topsoil in New York State to increase in thickness?
1   an increase in slope
2   a decrease in rainfall
3   an increase in biologic activity
4   a decrease in air temperature

Base your answers to questions 21 through 24 on your knowledge of Earth Science and on the vertical cross section showing a stream profile with reference points *A* through *F* within the stream bed.

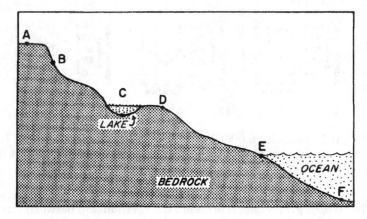

21 The primary force causing the movement of materials from point *B* to point *F* is
1   air pressure   2   insolation   3   water   4   gravity

22 At which point would erosion most likely be greatest?
1   *A*   2   *B*   3   *C*   4   *D*

23 At which point would particles of colloidal size most likely be deposited?
1   *A*   2   *B*   3   *C*   4   *D*

24 Which would most likely happen if the stream discharge between points *D* and *E* were to increase?
1   The average velocity of water would increase.
2   The amount of soil erosion would decrease.
3   The size of the particles carried in suspension would decrease.
4   The length of the stream would decrease.

The Rock Cycle — The Depositional Process

Vocabulary To Be Understood In Topic X

Colloid
Deposition
Dynamic Equilibrium
Erosional-Depositional Process
Graded Bedding

Horizontal, Vertical Sorting
Quiet Medium
Sediment Laden Flow
Settling Rate, Time
Sorting of Sediments

A. Deposition

Deposition is a part of the **erosional-depositional system**, in which sediments carried by a transporting agent dropped from the medium. **Deposition** is also called **sedimentation**. Since water is the predominant transporting agent, final deposition (often) occurs at the end of a stream, where the stream flows into a larger body of water, such as a lake or an ocean. Dissolved ionic minerals and colloids are released by the process of precipitation, and larger particles settle out of the transporting medium.

What factors affect the deposition of particles in a medium?

Factors

The major factors that affect the rate of deposition are particle size, shape and density, and the velocity of the transporting medium.

Size. The smaller particles settle more slowly than the larger particles, because as the velocity of a stream slows, the larger and usually heavier particles drop the fastest pulled down by gravity. The smaller particles tend to remain in suspension for longer periods of time, settling more slowly.

Colloids, particles approximately 10^{-4} to 10^{-6} millimeters across, generally remain in suspension indefinitely as long as there is even the slightest movement of the transporting medium.

When there is little or no movement in the transporting medium, sorting in a **quiet medium** (such as still-water or air) forms horizontal layers.

Sorting In A Quiet Medium

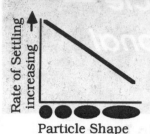

Rate of Settling increasing →

Particle Shape

Comparison Of Settling Rate And Particle Size

Shape. The other factors affecting deposition being equal, a round (spherical) shaped sediment particle will settle out of the transporting medium faster than a flat (disk) shaped particle. The greater the resistance of a flat particle the slower it will settle.

Density. Two particles of the same size and shape, but of different densities will settle at different rates. A high density particle will settle faster than a low density particle, since a high density particle is heavier.

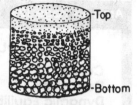

Top

Bottom

Rapid Deposition In A Quiet Medium

Velocity. As the velocity of a **sediment laden flow** decreases, the particles of greater weight and density settle out first. This decreasing velocity results in **horizontal sorting** (see the illustration at page bottom). That is, the heavier (more dense) particles settle first, and the lighter (less dense) particles settle on the stream bed farther downstream.

If deposition is rapid, **graded bedding**, or vertical sorting occurs. The heavier or more dense particles settle first, followed by the lighter or less dense particles. In general the bottom of the stream bed will have the larger particles and the particle size will decrease toward the top of the stream bed.

The particles in a moving medium do not necessarily move at the same velocity as the transporting medium. Particle movement is usually slower than the transporting medium movement.

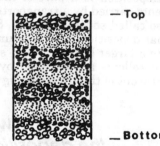

— Top

— Bottom

Graded Bedding In A Series Of Depositions

Colloidal particles in suspension tend to move at the same velocity as the stream, but particles that are more dense (heavier) or have a greater resistance (larger and flatter) often move at a slower velocity than the fluid transporting medium. Pebbles rolling along a stream bed travel slower than the water.

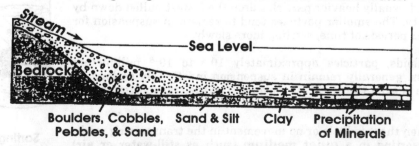

Stream →

Sea Level

Bedrock

Boulders, Cobbles,   Sand & Silt   Clay   Precipitation
Pebbles, & Sand   of Minerals

The cross section of the stream illustrates that larger, rounder, high density particles settle out of decreasing water velocity the fastest.

Sorting in a quiet, solid medium, such as the ice of a glacier, is more complex than in a fluid medium. The deposition is often unsorted, since sediments of different sizes, shapes and densities are all deposited together.

B. Erosional – Depositional System

An erosional – depositional system shows the interrelationships of the erosion process, the transportation agents and medium, the erosional materials, and the process of deposition.

Characteristics Of The Erosional – Depositional System

What are some characteristics of an erosional – depositional system?

Erosional – Depositional Change

The erosional and depositional processes produce characteristic changes which can be observed (see Topic XIV). In general, where erosion occurs, the materials surrounding the stream are removed and the stream cuts wider and/or deeper (for example, a river gorge). Where deposition occurs, materials are deposited, causing the stream to become shallower and/or wider. Often the stream change direction or splits into branches (for example, a river delta).

Dominant Process

Either the process of erosion or the process of deposition may be dominant in a particular location. If the slope of a stream is steep, the velocity is greater and the erosional process is the dominant process in that location. However, if the slope of a stream is more gentle, and the velocity is less, the depositional process is dominant. Erosional affects are usually greater at the source of a stream, and depositional affects are greater at the stream's mouth.

There are specific locations in a stream where the erosional process is dominant. For example, a stream's velocity is greater near the outside edge of a curve, or **meander**. Therefore, erosion is greater (dominant) on the outside bank of the stream's curve, and deposition is greater (dominant) on the inside, where the water velocity is slower.

Erosional – Depositional Interface

As previously discussed, erosion and deposition occur at various locations throughout the length of a stream. Therefore, interfaces between erosion and deposition can often be located. These interfaces are found midstream in meanders, where changes in stream velocity occur near the mouths of streams.

Dynamic Equilibrium

Where neither the process of erosion nor deposition is dominant, a state of **dynamic equilibrium** exists. This state of balance would be found midstream in a river meander.

Energy Relationships

The erosion phase of the erosional – depositional system results from a transfer from potential to kinetic energy (energy gain), and the depositional phase results from the reverse transfer (energy loss).

A stream's maximum potential energy is at its source (head or beginning). As a stream flows (from high elevation to low elevation), its potential energy is being transformed into kinetic energy, but some of this kinetic energy is lost to the environment through friction as heat energy. Eventually, the velocity of the stream decreases to zero at the mouth (end or outlet) of the stream and there is a net loss in the energy of the stream. In an erosional – depositional system, the total energy within the system decreases.

Skill Assessments

Base your answers to questions 1 through 5 on your knowledge of Earth Science, the Reference Tables, and the cross-sectional diagram. The diagram shows a sediment-laden stream entering the ocean. Points X and Y are in the stream and the ocean is divided into four zones A, B, C, and D.

1   What type of sediment sorting is shown in the diagram? Explain what causes this type of sorting.

2   In what zone would limestone most likely form?

3   Which zone would contain particles mostly in the range of 0.05 to 0.10 centimeter in diameter?

4   If point X were uplifted, how would the potential energy of the stream be affected?

5   A stream deposit contains particles that range in diameter from 2 to 4 centimeters. What sedimentary rock would be formed when this sediment is compacted and cemented together?

Questions for Topic X

1 If all the particles illustrated have the same mass and density, which particle will settle fastest in quiet water? [Assume settling takes place as shown by arrows.]

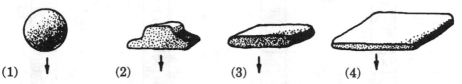

(1)　　　　　　　(2)　　　　　　　(3)　　　　　　　(4)

2 Why do particles carried by a river settle to the bottom as the river enters the ocean?
  1 The density of the ocean water is greater than the density of the river water.
  2 The kinetic energy of the particles increases as the particles enter the ocean.
  3 The velocity of the river water decreases as it enters the ocean.
  4 The large particles have a greater surface area than the small particles.

3 Which rock particles will remain suspended in water for the longest time?
  1 pebbles　　　　2 sand　　　　　　3 silt　　　　　　4 clay

4 In which location is erosion usually greater than deposition?
  1 in a stream channel that is being deepened
  2 along a coast where a sandbar is being enlarged
  3 at a point where a stream enters a lake
  4 at the base of a cliff where a pile of rock fragments is accumulating

Equipment used to measure settling times of the particles illustrated

Base your answers to questions 5 and 6 on the diagram and your knowledge of Earth Science.

Water

Particles of known size

2 meters

Supply of sorted particles

X　Y　Z

5 Which graph below best represents the effect of particle size on settling time if the particle densities are the same?

(1) SETTLING TIME / PARTICLE SIZE

(2) SETTLING TIME / PARTICLE SIZE

(3) SETTLING TIME / PARTICLE SIZE

(4) SETTLING TIME / PARTICLE SIZE

6   All three sediments in the diagram are placed in the cylinder with water, the mixture is shaken, and the particles are allowed to settle. Which diagram best represents the order in which the particles settled?

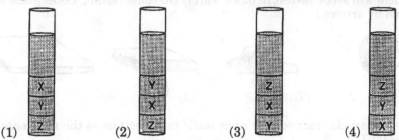

(1)   (2)   (3)   (4)

7   Sediments deposited by a glacier will most likely be
   1   horizontally sorted, only   3   horizontally and vertically sorted
   2   vertically sorted, only   4   unsorted

8   Small spheres that are identical in shape and size are composed of one of four different kinds of substances: *A, B, C,* or *D.* The spheres are mixed together and poured into a clear plastic tube filled with water. Which property of the spheres caused them to settle in the tube as shown in the diagram?

   1   their shape
   2   their size
   3   their density
   4   their hardness

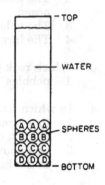

9   The diagram at the right represents a vertical cross section of sediments deposited in a stream.

   Which statement best explains the mixture of sediments?
   1   The velocity of the stream continually decreased/
   2   The stream discharge continually decreased.
   3   The particles have difference densities
   4   Smaller particles settle more slowly than larger particles.

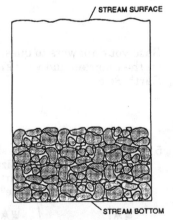

10   A dynamic equilibrium exists in an erosional–depositional system when
   1   all sediments are transported to the sea and erosion stops
   2   the amounts of the kinetic energy and potential energy both equal zero
   3   the rate of erosion exceeds the rate of deposition
   4   the rate of erosion is the same as the rate of deposition

11 The diagram below represents the cross section of
a soil deposit from a hill in central New York
State.
The deposition was most likely caused by

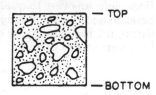

1   a glacier
2   a wind storm
3   a stream entering a lake
4   wave action along a beach

Base your answers to questions 12 through 14 on the diagram, the Reference
Tables, and your knowledge of Earth Science. The diagram represents a glacier
moving out of a mountain valley. The water from the melting glacier is flowing
into a lake. Letters *A* through *F* identify points within the
erosional/depositional system.

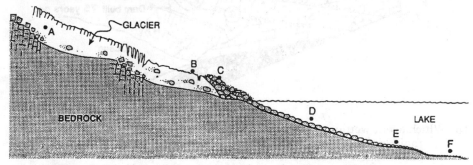

12 Deposits of unsorted sediments would probably be found at location
(1) *E*   (2) *F*   (3) *C*   (4) *D*

13 An interface between erosion and deposition by the ice is most likely located
between points
(1) *A* and *B*   (2) *B* and *C*   (3) *C* and *D*   (4) *D* and *E*

14 Colloidal-sized sediment particles carried by water are most probably being
deposited at point
(1) *F*   (2) *B*   (3) *C*   (4) *D*

15 Four samples of aluminum, *A*, *B*, *C*, and *D*, have
identical volumes and densities, but different
shapes. Each piece is dropped into a long tube
filled with water. The time each sample takes to
settle to the bottom of the tube is shown in the
table below.

Which diagram most likely represents the shape
of sample *A*?

Sample	Time to Settle (sec)
A	2.5
B	3.7
C	4.0
D	5.2

      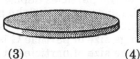

(1)   (2)   (3)   (4)

Base your answers to questions 16 through 20 on your knowledge of Earth Science and the block diagram of a portion of the Earth's surface. Numbers 1 through 5 indicate layers of earth material and letters *A* through *H* indicate locations on the surface.

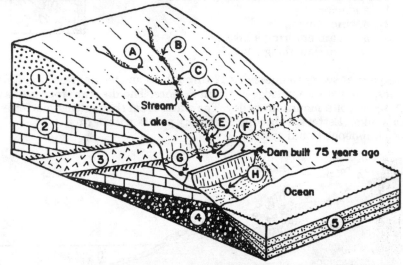

16 Which event would most likely convert the sediments in layer 5 into rock?
1 rapid cooling from the contact with the ocean
2 cementation of sediments caused by chemical processes
3 recrystallization due to heat from the intrusion of layer 3
4 heating and drying out due to rising convection currents

17 Compared to 75 years ago, why does the stream presently have less ability to downcut its channel at location *H*?
1 Much of the sediment used as tools for downcutting is being deposited behind the dam.
2 The energy of the stream is largely converted to heat from eroding the recently formed falls at location *E*.
3 Location *H* is now covered by a thick deposit of transported soil.
4 Humans have increased the discharge of water into the stream below the dam by large, sudden water releases.

18 At which location is the stream velocity probably the greatest?
1 A   2 F   3 C   4 E

19 Which earth material represented appears most resistant to weathering and erosion?
1 1   2 2   3 3   4 4

20 Particles from the stream are being deposited in the lake. How does the average size of the particles deposited beneath location F most likely compare to the average size of the particles deposited beneath location G?
1 The average size of particle F is greater.
2 The average size of particle G is greater.
3 The average size of particles at both locations is equal.

The Rock Cycle — The Formation Of Rocks

Vocabulary To Be Understood In Topic XI

Banding
Cementation
Compression
Crystal
Crystalline Structure
Distorted Structure
Evaporite
Extrusive Igneous Rock
Igneous Rock
Intrusive Igneous Rock
Metamorphic Rock
Minerals

Monomineralic
Nonsedimentary Rock
Polymineralic
Precipitation
Recrystallization
Rock Cycle
Rock-forming Mineral
Sedimentary Rock
Silicon – Oxygen Tetrahedron
Solidification (Crystallization)
Texture Characteristics
Transition Zone

A. Rocks And Sediments

The solid portion of the Earth's crust, the lithosphere, is composed of **naturally** formed material made up of one or more minerals, called **rock**. Much of the rock material on the surface of the Earth is composed of particles produced by weathering and erosion, called **sediment**.

The two major classifications of rocks are **sedimentary** and **nonsedimentary** (includes igneous and metamorphic) **rocks**.

Comparative Properties

What similarities do rocks have with sediments?

Similarities

Some rocks have properties that strongly resemble the sediments from which the rocks were formed. These properties include:

Discrete Layers
Different Sediments

• **Discrete layers** are separate and distinct layers of sediment, sorted during deposition, one on top of the other.

Organic (Fossil) Composition

• **Fragmental particles** (**clastics**) are pieces of sediment, cemented to form rock.

• **Organic** (**biogenic**) **composition** is rock composed of the remains of once living plants and/or animals.

• **A range of particle sizes**, often resembling a stream bed, delta, or other depositional area have larger more dense particles on the bottom and decreasing particle size and density towards the top of the rock.

• **A predominance of one particle size** indicate the deposition of one type, size, and density of sediment.

Predominance Of One Particle Size

Differences

Some rocks have properties that are very much different from their sediment origins. These include:

• **Various crystalline structures** often do not resemble the sediment minerals from which they developed. When high pressure and temperature cause the growth of mineral crystals (recrystallization), some of the rock's surrounding and original sediments are changed forming new crystals. This gives the rock new physical and/or chemical characteristics, called metamorphism.

Banding In A Metamorphic Rock

• **Banding of different minerals** gives the rock a layered appearance, involving the segregation of minerals in layers. This banding is the result of high pressure and heat. Generally, the more intense the temperature and pressure, the thicker the mineral bands that are formed. Thick banding indicates a high degree of metamorphism.

• **Distortion of structure** is the folding and bending of rock masses, brought about by very strong Earth forces, cause the rock to have a deformed structural appearance.

Distortion Of A Banded Rock By Pressure

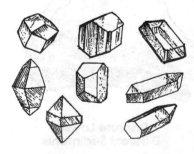

Minerals

Minerals are naturally occurring, crystalline, solid materials with definite chemical composition, molecular structure, and specific physical properties. Of the more than 2,000 minerals on the Earth, about a dozen of them are so abundant that they comprise more than 90% of the lithosphere. These common minerals are called the "**rock-formers**."

B. Mineral Relationship To Rocks

What is the composition of Rock?

Composition Of Rock

All rocks are composed of minerals. When only one mineral is found in a rock, the rock is said to be **monomineralic**. When a rock is composed of more than one mineral type, the rock is said to be **polymineralic**. Most rocks have a number of minerals in common.

Characteristics Of Minerals In Rocks

Minerals may be composed of single elements or compounds of two or more elements. Although there are more than 2,400 minerals, most of the Earth's crust is made of a relatively few common elements. The most abundant element is oxygen (O = 46.6% total Earth mass, 93.8% total Earth volume), and the second most abundant element is Silicon (Si = 27.7% total Earth mass, 0.9% total Earth volume). The other major elements are Aluminum (Al), Iron (Fe), Calcium (Ca), Sodium (Na), Potassium (K), and Magnesium (Mg).

What are some characteristics of minerals?

Physical And Chemical Properties Of Minerals

Minerals are identified on the basis of well-defined physical and chemical properties, such as color, hardness, streak, luster (surface reflecting qualities), fracture and cleavage planes (how the mineral breaks), structure (bonding of atoms), and taste (halite) and smell (sulfur).

Mineral Structure

The two most abundant elements in the Earth's crust are oxygen and silicon which unite to form a **tetrahedral** unit, the most common crystalline form. The **silicon – oxygen tetrahedron** can form crystals with itself, or combine with other elements to form different minerals with different external characteristics.

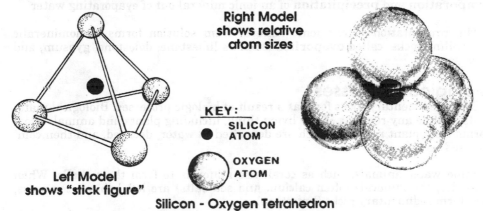

**Right Model
shows relative
atom sizes**

KEY:

● SILICON
ATOM

○ OXYGEN
ATOM

**Left Model
shows "stick figure"**

Silicon - Oxygen Tetrahedron

The internal structural arrangement of the atoms or tetrahedra give each mineral formed different physical properties, such as crystal shape, hardness, cleavage, and/or fracture.

C. Rock Formation

Rocks form as either sedimentary or nonsedimentary, depending on their composition and the environment in which the rock forms.

How are sedimentary rocks formed?

Sedimentary Rocks

As discussed in previous topics, the weathering processes produce sediments, which are transported by water, wind, and glaciers, and deposited in different locations on land or under water. Many of the sedimentary rocks form under large bodies of water in the following ways:

Compression, Cementation

Some sedimentary rocks form when the pressure of water and other overlying sediments compress very small particles (clay and colloids) that have settled from the transporting agent (deposition). The pressure may itself be sufficient to form these fine sediments into rock. An example of a compressed sedimentary rock is shale, although cementing may occur also.

Some sediments are combined with mineral cements that precipitate out of the ground water and result in cementation. Common mineral cements are iron, silica, and lime. Usually, cementation occurs with the larger sediments, such as sand, pebbles, and small rocks.

Sandstone is an example of a sedimentary rock in which the cemented particles are sorted in a uniform size. Most conglomerates are formed when the pebble sized particles are unsorted and cemented together. Frequently, the color of the rock is determined by the cementing agent.

Chemical Processes

Some sedimentary rocks form as a result of chemical processes, such as the **evaporation** and **precipitation** of an ionic mineral out of evaporating water.

The precipitation of one ionic mineral from solution forms monomineralic crystalline rocks, called **evaporites** (such as limestone, dolostone, gypsum, and rock salt).

Biological Processes

Some sedimentary rocks form as a result of biologic processes. Biological materials include any remains of any living thing, including plants and animals. Coal forms from plant remains which are deposited in water, decayed, and then compressed.

Some water animals, such as coral, use minerals to form their shells. When they die, the minerals (often calcium and sea salts) are left behind, compress, and form sedimentary rocks (limestone).

Scheme for Sedimentary Rock Identification

INORGANIC LAND-DERIVED SEDIMENTARY ROCKS					
TEXTURE	GRAIN SIZE	COMPOSITION	COMMENTS	ROCK NAME	MAP SYMBOL
Clastic (fragmental)	Mixed, silt to boulders (larger than 0.001 cm)	Mostly quartz, feldspar, and clay minerals; May contain fragments of other rocks and minerals	Rounded fragments	Conglomerate	
			Angular fragments	Breccia	
	Sand (0.006 to 0.2 cm)		Fine to coarse	Sandstone	
	Silt (0.0004 to 0.006 cm)		Very fine grain	Siltstone	
	Clay (less than 0.0006 cm)		Compact; may split easily	Shale	

CHEMICALLY AND/OR ORGANICALLY FORMED SEDIMENTARY ROCKS					
TEXTURE	GRAIN SIZE	COMPOSITION	COMMENTS	ROCK NAME	MAP SYMBOL
Nonclastic	Coarse to fine	Calcite	Crystals from chemical precipitates and evaporites	Chemical Limestone	
	Varied	Halite		Rock Salt	
	Varied	Gypsum		Rock Gypsum	
	Varied	Dolomite		Dolostone	
	Microscopic to coarse	Calcite	Cemented shells, shell fragments, and skeletal remains	Fossil Limestone	
	Varied	Carbon	Black and nonporous	Bituminous Coal	

How are nonsedimentary rocks formed?

Nonsedimentary Rocks

There are two main classifications of nonsedimentary rocks: **igneous** and **metamorphic**. These nonsedimentary rocks are formed from the solidification and crystallization of molten rock from within the Earth's surface and by the recrystallization of other rocks, respectively.

Solidification Process

Beneath the Earth's surface, molten rock material is called **magma**. When the magma reaches the Earth's surface, it is called **lava**. From this material, nonsedimentary rocks form.

When magma or lava cools and solidifies, it forms **igneous rock**. As the liquid rock solidifies and becomes hard, mineral crystals form, resulting in the igneous rock having a **crystalline texture**. Usually there are many different minerals (polymineralic) within this kind of rock.

The size of the crystals vary according to the conditions of time, temperature, and the pressure under which they form. In order for crystals to form (crystallization) the molten material must cool. The longer it takes for the molten material to cool, the larger the crystals that form and the more coarse the rock's texture.

The **texture** of the igneous rocks is dependent upon the rate of cooling. Slow cooling produces large crystals, rocks with a coarse texture (phaneritic), and rapid cooling produces small crystals, rocks with a fine texture (aphanitic).

Cooling is related to both temperature and pressure. Rapid cooling produces small crystals or a grainy texture. Very rapid cooling results in no crystals or a glassy texture. This is the result of a rapid drop in temperature or a pressure decrease. These conditions would exist near the surface of the Earth or when the molten material (lava) breaks through the surface, as in a volcano. When lava flows out and hardens on the Earth's surface, it is called an **extrusion**, and the rock formed is called, **extrusive igneous rock**.

However, deep within the Earth both temperature and pressure is much higher than at the surface. Therefore, cooling is slower and crystal size is larger.

Scheme for Igneous Rock Identification

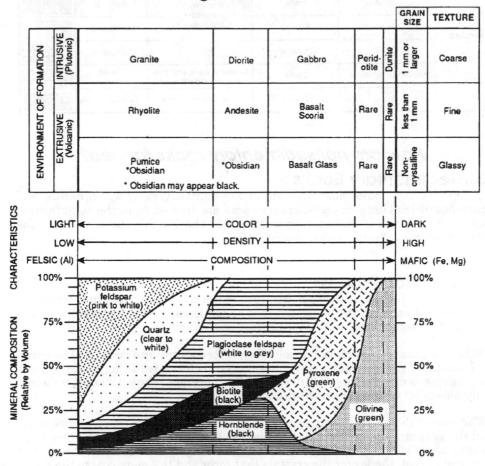

Note: The intrusive rocks can also occur as exceptionally coarse-grained rock, Pegmatite.

Scheme for Metamorphic Rock Identification

TEXTURE		GRAIN SIZE	COMPOSITION	TYPE OF METAMORPHISM	COMMENTS	ROCK NAME	MAP SYMBOL
FOLIATED	Slaty	Fine	CHLORITE MICA QUARTZ FELDSPAR AMPHIBOLE GARNET PYROXENE	Regional	Low-grade metamorphism of shale	Slate	
	Schistose	Medium to coarse			Medium-grade metamorphism; Mica crystals visible from metamorphism of feldspars and clay minerals	Schist	
	Gneissic	Coarse		(Heat and pressure increase with depth, folding, and faulting)	High-grade metamorphism; Mica has changed to feldspar	Gneiss	
NONFOLIATED		Fine	Carbonaceous		Metamorphism of plant remains and bituminous coal	Anthracite Coal	
		Coarse	Depends on conglomerate composition		Pebbles may be distorted or stretched; Often breaks through pebbles	Meta-conglomerate	
		Fine to coarse	Quartz	Thermal (including contact) or Regional	Metamorphism of sandstone	Quartzite	
			Calcite, Dolomite		Metamorphism of limestone or dolostone	Marble	
		Fine	Quartz, Plagioclase	Contact	Metamorphism of various rocks by contact with magma or lava	Hornfels	

When magma hardens in the Earth, it is called an **intrusion**. The rock formed is called **intrusive igneous rock**.

Recrystallization Process

Metamorphic rocks are igneous, sedimentary, and metamorphic rocks that have been changed in form usually deep within the Earth. The high pressures, temperatures, and chemical solutions deep in the Earth cause changes in the existing rocks, forming metamorphic rocks.

Metamorphism is the result of the **recrystallization** of *unmelted* material under high temperature and pressure. As previously explained, these extreme conditions cause the mineral crystals to grow and new minerals to form without melting (a solid state reaction). The weathering and erosion of the Earth's surface eventually exposes the metamorphic rocks.

D. Environment Of Formation

The composition, structure, and texture of a rock is dependent upon the environment in which the rock was formed.

What is the environment in which a rock forms?

Inferred Characteristics

The type of environment in which a rock was formed can be inferred from the rock's characteristics of composition, structure, and texture.

Composition. If a sedimentary rock's composition is limestone, the rock was probably formed through the processes of evaporation and/or precipitation of

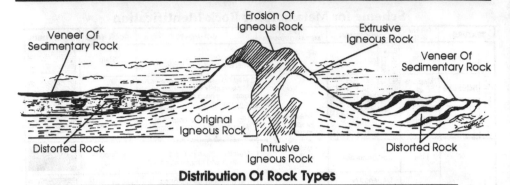

Distribution Of Rock Types

limewater solution. Large deposits of salt found beneath the Earth's surface, indicate that the rock was formed from the precipitation of the mineral halite from a large body of salt water and later buried under layers of other sediment. Such is the case of the salt mined from underground sources in central New York along Lake Seneca. The presence of marine animal fossils indicates that the rock formed near the Earth's surface by precipitation from an ocean.

Structure. Igneous rocks containing very large crystals formed slowly deep within the Earth's lithosphere. Metamorphic rocks that show evidence of distortion and banding were formed in an environment characterized by extreme pressure.

Texture. Since the texture of a rock is the direct result of size, shape, and arrangement of its mineral crystals, it is possible to determine the rock's origin. An igneous rock containing fine grains and having a glassy texture probably formed from the extrusion of molten liquid. Whereas, rocks containing uniformly large crystals, a coarse texture, probably formed deep within the Earth.

The texture of sedimentary rock may indicate the source and conditions under which the sediments were deposited. For example, the "frosted" grains of wind deposited sandstone are more pitted than the smoother grains in water deposited sandstone.

Distribution

Sedimentary rocks are usually found as a thin layer, or veneer, over large areas of continents. Nonsedimentary rocks, at or near the surface, are found in regions of volcanoes or mountains where intrusion and extrusion have occurred. Where intrusive igneous and metamorphic rocks are found exposed on the surface of the Earth, it is usually due to removing of over-lying rock through millions of years of weathering and erosion by water, wind, and glaciers.

E. The Rock Cycle

The rock cycle is a model to show how closely the rock types are related. Generally, the nonsedimentary rock type igneous is considered to be the primary or parent rock of the Earth's crust. As rocks are attacked by the forces of weathering and erosion, the sediments formed are deposited, cemented, and sedimentary rocks are formed.

Rock Cycle in Earth's Crust

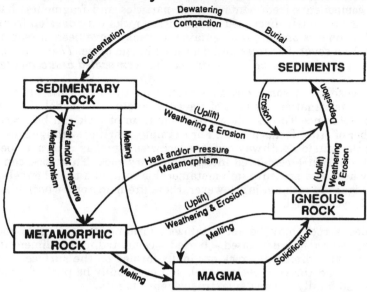

If these rocks are exposed to the pressure and temperature extremes, they may be changed into metamorphic rocks. These rocks may then be subjected to additional Earth forces, melting them into magma and eventually solidifying into igneous rock. The rock cycle is then completed.

Any type rock, igneous, sedimentary, or metamorphic, may be changed into any other type depending upon the environment to which it is subjected. There is no preferred or predictable path that a rock will take within the environment.

Evidence For The Rock Cycle

What evidence suggests a cycle model of rock formation?

Transition Zones. Where molten magma or lava has come in contact with other rocks and has changed them, no specific point exists between changed and unchanged rock. A transition zone may form having characteristics of both rock types, due to the gradual blending of the rock types.

Any rock can be changed by the action of increased pressure, temperature, and/or contact with molten material passing through the local rock.

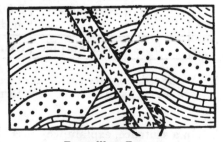

Transition Zones
(boundary between sedimentary rock and intrusive igneous rock)

Rock Composition. The composition of some sedimentary rocks suggests that the components (sediments or rock particles and fragments) had **varied origins**. As previously discussed, sedimentary rocks may develop from a combination of sediments and organic remains that may have been transported great distances from several sources and deposited in one place. *Then, the forces of the environment change these varied sediments into compacted and cemented rock.*

The composition of some rocks suggests that the materials have undergone multiple transformations as part of the **rock cycle**. The model of the rock cycle (from the Reference Tables) shows how one form of rock can be processed into any number of other forms of rock. For example, starting with igneous rocks and following the outside pathway counterclockwise, igneous rock may be eroded – deposited – buried – cemented into sedimentary rock. That rock can be transformed by heat and pressure into metamorphic rock, and then be melted and solidify into igneous rock again. However, there are many other possible pathways in the rock cycle model.

The igneous rock may be recrystallized, forming metamorphic rock, which could then be eroded – deposited – buried – cemented into sedimentary rock. It is possible that sedimentary rock originally formed on the surface, could take a pathway through the rock cycle model, and eventually be part of a molten rock extrusion and finally become an extrusive igneous rock.

Skill Assessments

Base your answers to questions 1-4 on your knowledge of Earth Science, the Reference Tables and the data below for five different rock samples.

Data Table

ROCK SAMPLE	ORIGIN	CRYSTAL SIZE OR GRAIN SIZE	OTHER CHARACTERISTIC
1	igneous	no crystals	glassy
2	igneous	coarse	light color
3	igneous	fine	dark color
4	sedimentary	0.0003 cm in diameter	contains dinosaur footprints
5	metamorphic	coarse	shows banding

1　What sedimentary rock is represented by sample 4? During what geologic era was it likely to have been formed?

2　In a sentence, explain why rock sample one has a glassy texture.

3   In a short paragraph, explain the probable cause of the banding
    characteristic seen in rock sample 5.

4   Which rock sample is granite? Explain how you know.

5   Devise a classification system for the following rocks:
    Salt, conglomerate, shale, bituminous coal, gypsum, fossil limestone,
    sandstone, chemical limestone

6   In complete sentences, describe 5 ways in which metamorphism may have
    altered a parent rock.

Questions For Topic XI

1   What causes the characteristic crystal shape
    and cleavage of the mineral halite as shown in
    the diagram at the right?
    1   metamorphism of the halite
    2   the internal arrangement of the atoms in
        halite
    3   the amount of erosion the halite has
        undergone
    4   the shape of other minerals located where
        the halite formed

HALITE
(salt)

2   Why do diamond and graphite have different physical properties, even
    though they are both composed entirely of the element carbon?
    1   Only diamond contains radioactive carbon.
    2   Only graphite consists of organic material.
    3   The minerals have different arrangements of carbon atoms.
    4   The minerals have undergone different amounts of weathering.

3   Which mineral is commonly found in granite?
    1   quartz   3   magnetite
    2   olivine   4   galena

4   Which element combines with silicon to form the tetrahedral unit of
    structure of the silicate minerals?
    1   oxygen   3   potassium
    2   nitrogen   4   hydrogen

5   The data table at the right shows the composition of six common rock-forming mineral.

Mineral	Composition
Mica	$KAl_3Si_3O_{10}$
Olivine	$(FeMg)_2SiO_4$
Orthoclase	$KAlSi_3O_8$
Plagioclase	$NaAlSi_3O_8$
Pyroxene	$CaMgSi_2O_6$
Quartz	SiO_2

The data table provides evidence that

1   the same elements are found in all minerals
2   a few elements are found in many minerals
3   all elements are found in only a few minerals
4   all elements are found in all minerals

6   Which object is the best model of the shape of a silicon-oxygen structural unit?

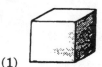

(1)   (2)   (3)   (4)

7   When dilute hydrochloric acid is placed on the sedimentary rock limestone and the nonsedimentary rock marble, a bubbling reaction occurs with both. What would this indicate?
1   The minerals of these two rocks have similar chemical composition.
2   The molecular structures of these two rocks have been changed by heat and pressure.
3   The physical properties of these two rocks are identical.
4   The two rocks originated at the same location.

8   Two mineral samples have different physical properties, but each contains silicate tetrahedrons as its basic structural unit. Which statement about the two mineral samples must be true?
1   They have the same density.
2   They are similar in appearance.
3   They contain silicon and oxygen.
4   They are the same mineral.

9   Which statement best describes a general property of rocks?
1   Most rocks have a number of minerals in common.
2   Most rocks are composed of a single mineral.
3   All rocks contain fossils.
4   All rocks contain minerals formed by compression and cementation.

10 According to the Reference Tables, which property would be most useful for identifying igneous rocks?
1   kind of cement   3   number of mineral present
2   mineral composition   4   types of fossils present

11 The diagram represents the percentage by volume of each mineral found in a sample of basalt.

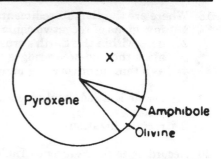

Which mineral is represented by the letter X in the diagram?
1 orthoclase feldspar
2 plagioclase feldspar
3 quartz
4 mica

12 According to the Reference Tables, rhyolite and granite are alike in that they both are
1 fine-grained
2 dark-colored
3 mafic
4 felsic

13 According to the Reference Tables, which graph best represents the comparison of the average grain sizes in basalt, granite, and rhyolite?

Key to Graph Abbreviations
B – Basalt
G – Granite
R – Ryolite

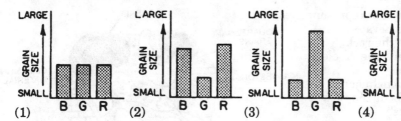

14 Which characteristic of an igneous rock would provide the most information about the environment in which the rock solidified?
1 color
2 texture
3 hardness
4 streak

15 Which is the best description of the properties of basalt?
1 fine-grained and mafic
2 fine-grained and felsic
3 coarse-grained and mafic
4 coarse-grained and felsic

16 Large crystal grains in an igneous rock indicate that the rock was formed
1 near the surface
2 under low pressure
3 at a low temperature
4 over a long period of time

17 The diagram shows an igneous intrusion in sedimentary rock layers.

At which point would metamorphic rock most likely be found?
(1) A
(2) B
(3) C
(4) D

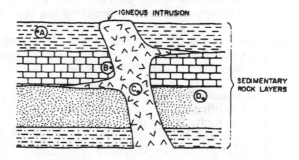

18 Where are the Earth's sedimentary rocks generally found?
1 in regions of recent volcanic activity
2 deep within the Earth's crust
3 along the mid-ocean ridges
4 as a thin layer covering much of the continents

19 Limestone is a sedimentary rock which may form as a result of
1 melting   3 metamorphism
2 recrystallization   4 biologic processes

20 According to the Reference Tables, which sedimentary rock most likely formed as an evaporite?
1 siltstone   3 gypsum
2 conglomerate   4 shale

21 What rock is formed by the compression and cementation of sediments with particle sizes ranging from 0.08 to 0.1 centimeter?
1 basalt   3 granite
2 conglomerate   4 sandstone

22 The diagram represents a conglomerate rock. Some of the rock fragments are labeled.

Which conclusion is best made about the rock particles?

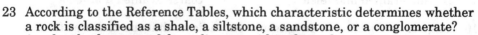

1 They are the same age.
2 They originated from a larger mass of igneous rock.
3 They all contain the same minerals.
4 They have different origins.

23 According to the Reference Tables, which characteristic determines whether a rock is classified as a shale, a siltstone, a sandstone, or a conglomerate?
1 the absolute age of the sediments within the rock
2 the mineral composition of the sediments within the rock
3 the particle size of the sediments within the rock
4 the density of the sediments within the rock

24 Most of the surface bedrock of New York State formed as a direct result of
1 volcanic activity   3 melting and solidification
2 spreading of the ocean floor   4 compaction and cementation

25 Some nonsedimentary rocks are formed as a result of
1 solidification of molten material
2 evaporation and precipitation
3 cementation of particles
4 deposition of particles

Base your answers to questions 26 through 30 on your knowledge of Earth Science, the Earth Science Reference Tables, and the diagram below. The diagram represents a geologic cross section consisting of various sedimentary and non sedimentary rocks which have not been overturned.

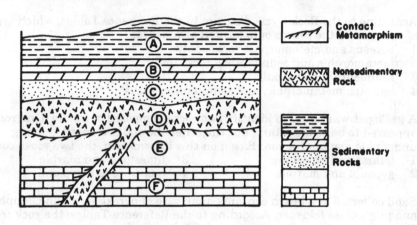

26 Fossils would least likely be found in rock layer
1 *A*   2 *B*   3 *C*   4 *D*

27 What can be inferred about the relative age of the various rocks?
1 Rock *C* is older than rock *E*.
2 Rock *C* is older than rock *D*, but younger than rocks *A* and *B*.
3 Rock *D* is older than rock *C*, but younger than rock *E*.
4 Rock *D* is older than rocks *C, E*, and *F*

28 Rock layer *A* could have formed by the
1 deposition of clay   3 rapid cooling of molten material
2 metamorphism of slate   4 recrystallization of basalt

29 Rock layer *E* is composed of nonuniform particle sizes ranging in diameter from 0.9 to 23 centimeters. According to the Reference Tables, this rock layer should be represented by which symbol?

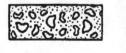

(1)   (2)   (3)   (4)

30 Rock layer *D* is classified as nonsedimentary because it was formed by
1 the compression of particles   3 the evaporation of seawater
2 the cooling of molten material   4 biologic processes

31 Metamorphic rocks result from the
1 erosion of rocks
2 recrystallization of rocks
3 cooling and solidification of molten magma
4 compression and cementation of soil particles

32 Which rocks would most likely be separated by a transition zone of altered rock (metamorphic rock)?
   1   sandstone and limestone
   2   granite and limestone
   3   shale and sandstone
   4   conglomerate and siltstone

33 According to the Rock Cycle diagram in the Reference Tables, which type(s) of rock can be the source of deposited sediments?
   1   igneous and metamorphic rocks, only
   2   metamorphic and sedimentary rocks, only
   3   sedimentary rocks, only
   4   igneous, metamorphic, and sedimentary rocks

34 A geologist was asked to identify two rocks composed of calcite. One rock appeared to be sedimentary, but the other showed evidence of having undergone metamorphism. Based on this information, the two rocks could be
   1   quartz and limestone
   2   gypsum and marble
   3   limestone and marble
   4   quartzite and gypsum

35 Sand collected at a beach contains a mixture of pyroxene, olivine, amphibole, and plagioclase feldspar. According to the Reference Tables, the rock from which this mixture of sand came is best described as
   1   dark-colored with a mafic composition
   2   dark-colored with a felsic composition
   3   light-colored with a mafic composition
   4   light-colored with a felsic composition

36 The green sand found on some Hawaiian Island shorelines most probably consists primarily of
   1   quartz
   2   olivine
   3   plagioclase feldspar
   4   orthoclase feldspar

The Rock Cycle — The Dynamic Crust

Vocabulary To Be Understood In Topic XII

Bench Mark
Compressional Wave (P-wave)
Continental Crust, Drift
Crust
Earthquake
Epicenter
Fault
Focus of an Earthquake
Folded Strata
Fossils
Geosyncline
Inner Core
Isostasy

Mantle
Mantle Convection Cells
Mid-Ocean Ridge
Oceanic Crust
Outer Core
Plate Tectonics
Reversal of Magnetic Polarity
Sea (Ocean) – Floor Spreading
Seismic Waves, Seismograph
Shear Wave (Secondary, S-wave)
Strata
Subsidence
Tilted Strata

A. Evidence For Crustal Movement

The solid rock outer zone of the Earth is known as the **lithosphere** or **crust**. This crust is in a constant state of change, and there is much evidence to support the idea that the Earth's surface has always been changing.

Some of these changes can be directly observed, such as the results of earthquakes, crustal movements (both horizontal and vertical) along fault zones. and volcanoes on the surface of the Earth. Other evidence indicates that parts of the Earth's crust have been moving to different locations for billions of years.

These and other evidences for crustal movement are divided into minor and major crustal changes.

Minor Crustal Changes

The minor crustal changes include the deformation and displacement of strata and fossils.

What evidence suggests minor changes in the Earth's crust?

Deformed Rock Strata

Sedimentary rocks form in horizontal layers, but observations of the Earth's surface indicate that the original formations of rock were changed through past Earth movements. These changes include **tilting**, **folding**, **faulting**, and the **displacement** of these layers of rock or **strata**.

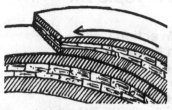

Thrust Faulting

Displaced Fossils. Marine fossils, the remains or imprints of once living ocean organisms, such as corals, fish, and other marine animals, are found in layers of sedimentary rock in mountains, often thousands of feet above sea level. These marine fossils found at high elevations above sea level suggest the past **uplift of rock strata**.

Subsidence is the sinking or settling of rock strata. Observations at great depths in the oceans suggest that subsidence has occurred in the past, since fossils of shallow water organisms, marine animals that once lived near the ocean surface, have been found in the rocks of the ocean floor.

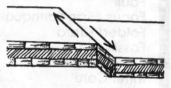

Horizontal Faulting

Displaced Strata. The displacement of strata provides direct evidence of crustal movement. These crustal movements may be either horizontal or vertical or a combination of both.

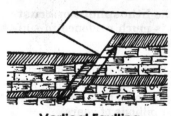

Vertical Faulting

Horizontal Displacement (faulting) occurs when the Earth's surface shifts sideways along a transform fault or crack in the crust. **Vertical displacement** (faulting) occurs when a portion of the Earth's surface is either uplifted or subsides, also along a fault. For example, in California, where earthquakes are fairly common crustal activity, both types of displacement can be observed along the San Andreas Fault Zone.

During the San Fernando Earthquake of 1971, the horizontal displacement along the fault zones caused portions of roads to be displaced so that one road was separated into two. Because of the uplift and subsidence of adjacent land, cliffs and ravines were formed. The actual amount of movement can be calculated by comparing bench marks.

A **bench mark** is a permanent cement or brass marker in the ground indicating elevation above mean sea level and the latitude and longitude of that particular location. Following the earthquake some bench marks were displaced several meters vertically and horizontally.

Major Crustal Changes

The major crustal changes include geosynclines, vertical crustal movements, plate tectonics (the spreading of the ocean floor away from the mid-ocean ridges and continental drift), and changes in the Earth's magnetic poles.

What evidence suggests major changes in the Earth's crust?

Zones Of Crustal Activity

The major form of crustal activity on the Earth today includes shifts (deformations) in the Earth's crust. Activities associated with mountain building processes include earthquakes, geosynclines, and volcanoes.

These major crustal activities occur for the most part in specific zones, or regions, of the the Earth. They are located at the boundaries of crustal plates (discussed later in this topic). Generally, they are along the borders of continents and oceans. The most active zones generally follows the continental borders of the Pacific Ocean, mid-ocean ridges, and across southern Europe and the Middle East into Asia.

Geosynclines

A shallow ocean basin along the edge of a continent that is sinking is called a **geosyncline**. A geosyncline forms as a result of millions of years of sedimentary deposition that builds thick layers of sediment in large regions where there is shallow water. As the weight of the sediments increases, subsidence occurs. The bottom of the geosyncline slowly sinks, making room for more sediment in the basin.

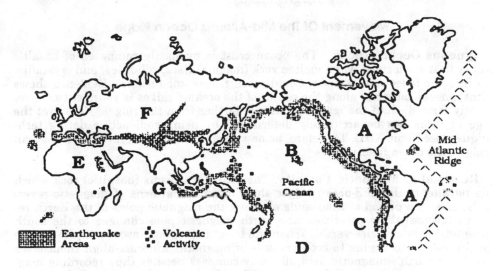

Zones of Crustal Activity On The Earth
Major Plate Areas: A - American, B - Pacific, C - Nazca, D - Antarctic,
E - African, F - Eurasian, and G - Indian

This *balancing* between greater sedimentary deposition and more room for the sediments is an example of **isostasy**. Basically, the Earth's crust is in a state of equilibrium with any change in one part of the crust being offset by an change in another part of the crust.

Geosynclines are associated with the enlargement of continents by accretion and mountain building. The thick sedimentary rocks are affected by Earth forces causing the region to be uplifted, forming mountains. Other geosyncline areas may be slowly subsiding. The Gulf of Mexico may be a present-day geosyncline.

Vertical Movements

Raised beaches or terraces and changed bench mark elevations can be observed and are the result of major earth movements. Vertical movements can be related to things such as plate movement and isostatic compensation.

Sea (Ocean) Floor Spreading

There is much evidence to indicate that the ocean floors are spreading out from the mountain regions of the oceans (mid-ocean ridges). The two major evidences are related to the **age of igneous ocean materials** and the **reversal of magnetic polarity**.

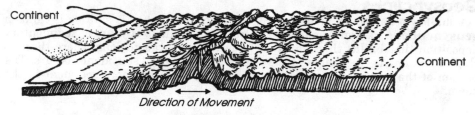

Continent

Continent

Direction of Movement

Movement Of The Mid-Atlantic Ocean Ridge

Igneous Ocean Rocks. The ocean crust is primarily composed of basaltic rocks that are formed when molten rock (magma) rises, solidifies, and crystallizes into the igneous rocks of the mid-ocean ridges (mountains). Evidence shows that igneous material along the center of the oceanic ridges is younger (more recently formed) than the igneous material farther from the ridges. Note that the age of rocks can be accurately determined by using radioactive dating techniques (see Topic XIII). Therefore, as new ocean crust is generated at mid-ocean ridges, the ocean floor widens.

Reversal of Magnetic Polarity. The strips of igneous (basaltic) rock which lie parallel to the mid-ocean ridges show matched patterns of magnetic reversals. Over the period of thousands of years, the magnetic poles of the Earth reverse (switch) their polarities (the north magnetic pole changes to the south magnetic pole and vice versa). When the basaltic magma flows up in the middle of the ridge and begins to cool, crystals of magnetic minerals align themselves with the Earth's magnetic field, like tiny compass needles thus recording magnetic polarity at the time of formation. Therefore, when the magnetic field is reversed, the new igneous rocks formed during the reversed polarity period have a reversed magnetic orientation from the previously formed rocks. These changes in magnetic orientation are found on both sides of the mid-ocean ridges, indicating that the development of the ocean floor is from the mid-ocean ridges outwards.

Continental Drift

The concept that the major land masses, the continents, are moving around on the Earth's surface is known as **continental drift**. The evidence supporting the concept of continental drift is both direct and indirect.

Continental Outlines. The present continents appear to fit together as fragments of an originally larger land mass, much in the same way as the pieces of a picture puzzle fit together. The best evidence indicates that approximately 200 million years ago, the major continents were connected, and that since then the continents have been moving generally apart (see Topic XIII and the Reference Tables, Inferred Positions of the Earth Landmasses).

Rocks, Minerals and **Fossils.** The correlation of rock, mineral, and fossil evidence between continents suggests that the land masses were joined at some time in the past. For example, many of the same rock types and their mineral composition, as well as the fossil types found in those rocks along the eastern coastline of South America match those along the western coastline of Africa.

Additional Evidence. The alignment of crystals and the magnetic orientations of rock types found at the opposite edges of continents, indicate that these rocks were formed under the same conditions and at the same time in geologic history. A comparison of plant and animal life forms, both in past (fossils) and present (living) history, shows similarities of common origin, thus evidence that these organisms developed at the same location and time. Although some migration of animals and transportation of plant seeds did occur, commonalities in terrestrial animals on presently widespread continents would seem to indicate that the continents were probably connected in past history.

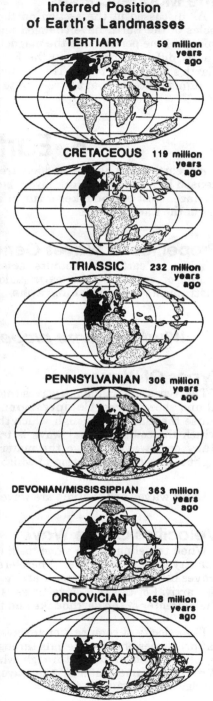

Inferred Position of Earth's Landmasses

TERTIARY — 59 million years ago

CRETACEOUS — 119 million years ago

TRIASSIC — 232 million years ago

PENNSYLVANIAN — 306 million years ago

DEVONIAN/MISSISSIPPIAN — 363 million years ago

ORDOVICIAN — 458 million years ago

The Magnetic Poles

As discussed in the section on the sea (ocean) floor spreading, the magnetic poles of the Earth have changed magnetic polarity. The rocks formed during a specific time period have the magnetic orientation of that period. Rocks found in vastly different locations in the Earth's crust have recorded the position of the Earth's magnetic poles. This supports the continental growth and mountain building concepts that are related to plate tectonics.

B. Earthquakes

An earthquake is the sudden trembling or shaking of the ground, usually caused by a shifting of rock layers along a fault or fissure under or at the Earth's surface. Most earthquakes occur in zones of crustal activity, along the boundaries of the Earth's plates.

Properties Of Waves Generated By An Earthquake

When an earthquake occurs, **seismic waves** are generated from the **focus** (point of origin) of the earthquake and are registered on delicate sensing instruments called **seismographs**.

What are some properties of earthquake waves?

Types Of Waves

Two types of earthquake waves that travel through the Earth are compressional and shear waves. **Compressional waves**, also called **Primary** or **P-waves**, cause the material through which they pass to vibrate in the same direction in which the compressional wave is traveling. **Shear waves**, also called **Secondary** or **S-waves**, cause the material through which they pass to vibrate at right angles to the direction in which the shear wave is traveling.

Long waves or **L-waves** are waves that travel along the surface of the Earth.

Velocities Of The Waves

When traveling in the same medium, compressional waves travel at a greater velocity than shear waves. Therefore, a seismograph records the compressional waves before the shear waves arrive. A comparison of the time recorded between the arrival of the two wave types is used in determining the distance between the epicenter of the earthquake and the recording station (seismograph).

The velocity of compressional waves and shear waves depends upon the physical properties of the materials through which they travel. The greater the density of the solid material through which they travel, the greater their velocity. Within the same material, the waves will have a greater velocity if there is an increase in pressure.

Transmission Of Waves

Compressional waves are transmitted through both solids, liquids, and gases, but shear waves are only transmitted through solids. This difference provides valuable information for scientists about the structure of the Earth's interior.

Location Of An Epicenter

The **epicenter** of an earthquake is the point on the Earth's surface that is directly above the focus of the earthquake.

How can an earthquake epicenter be located?

Epicenter

Compressional and Shear waves travel at different velocities through the same medium with compressional waves traveling faster than shear waves. Differences in the arrival times of the seismic waves can be used to determine the distance to the epicenter from the seismographic station. The larger the difference between the arrival times of the compressional and shear waves, the greater the distance from the recording station. By using the distance to the epicenter from three different seismographic stations, the location of the earthquake's epicenter can be determined.

Origin Of The Time Of An Earthquake

The origin time of an earthquake can be inferred from the evidence of the epicenter distance and the travel time of the P-waves and S-waves (see the Earthquake S-wave and P-wave Time Travel Graph below). The farther the recording station is from the epicenter, the longer it takes for the P-waves to reach the station. For example, a recording station receives the P-wave at 7:10 a.m. and the S-wave 5 minutes and 30 seconds later. If the epicenter distance has been determined to be 4,000 kilometers (determined by the difference - 5.5 minutes- in the arrival time of the P- and S-waves), then the travel time for the P-wave was 7 minutes. The earthquake's time of origin was 7:03 a.m. (7:10 less 07 minutes P-wave travel time).

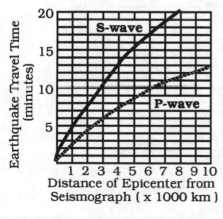

Distance of Epicenter from
Seismograph (x 1000 km)

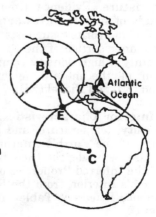

Locating An
Epicenter
Using Three
Seismograph
Stations
A, B, and C

The Epicenter
is located at
the intersection
of the three
distances,
marked E.

C. Model Of The Earth's Crust And Interior

Properties

What are some properties of the Earth's crust and interior?

Solid and Liquid Zones. Analysis of seismic data leads to the inference that solid and liquid zones exist within the Earth. There are four major Earth zones, three solid zones, and one liquid zone. The **crust**, the **mantle**, and the **inner core** are solid zones. The only liquid zone is the **outer core**.

Crustal Thickness. The crust of the Earth compared to the other zones is relatively thin, only a few kilometers in average depth. The average thickness of the **continental crust** is greater than the average thickness of the **oceanic crust** (ocean area).

Crustal and Interior Composition. The oceanic and continental crusts have different compositions. The continental crust of the Earth is mainly composed of a low density felsic granitic material whereas the ocean crust is mafic or basaltic in nature. Evidence from the study of seismic waves and metallic meteorites suggests that the inner portion of the Earth is mainly a high density combination of the metallic elements iron (Fe) and Nickel (Ni).

Interior Characteristics. The density, temperature, and pressure of the Earth's interior increase with depth. (See figure on the Inferred Properties of the Earth's Interior, from the Earth Science Reference Tables at the right.)

Inferred Properties of Earth's Interior

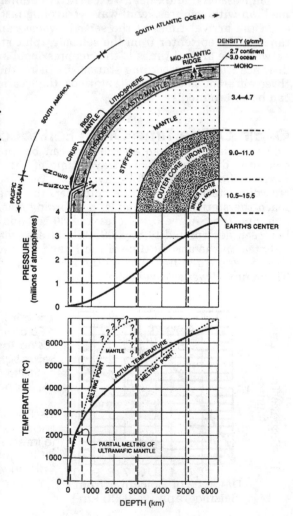

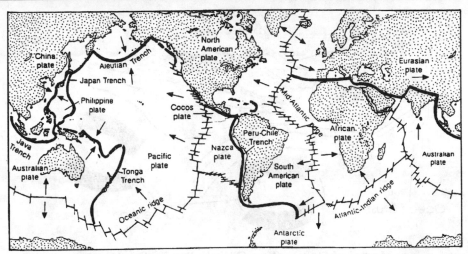

Plate Tectonics is the theory that the Earth's crust is made of solid lithospheric plates that move in relation to each other. This world map identifies the major plates, trenches, and ridges that are sites of major crustal activity.

D. Crustal Change (Theories)

The Earth's crust has changed in the past and is still in a constant state of change. The processes of Earth change have been discussed in this and the previous topics.

Inferred Processes

There are several theories, supported by both direct and indirect evidence, that suggest why these changes occur.

What inferences can be drawn about the processes which may cause crustal changes?

Plate Tectonics

The term tectonics refers to the forces that deform the Earth's crust. **Plate tectonics** is the theory that the Earth's crust is made of a number of solid pieces, called plates, and these lithospheric plates move in relation to each other. Where plates move apart, magma flows up and new crust is formed. This occurs at ocean ridges. Where plates collide, the overriding plate may crumble up to form mountains while the subducting plate plunges down and is melted. Plate motion is only a few centimeters per year.

Mantle Convection Cells

A convection cell is a moving stream of heated material that is moving due to density differences. It is suggested

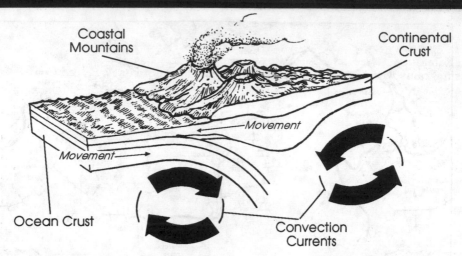

Model of Convection Cells in the Earth's Interior
Note: Not drawn to scale. Convection currents are believed to be large.

that convection cells exist within the mantle, because of the occurrence of heat flow highs in areas of mountain building, and heat flow lows in areas of shallow subsiding basins. The convection cells may be part of the mechanism which causes the continents to drift.

Geosynclinal Development

Continental growth and mountain building may be related to geosynclinal development as discussed in the section on evidences of major crustal changes (see page 151 in this Topic).

Isostasy

Mountains of geosynclinal origin may in part be caused by isostatic adjustments (vertical movement) of material of different density. Isostasy is the condition of equilibrium or balance in the Earth's crust. Since the upper mantle acts like a fluid ("plastic-like"), the crustal plates are floating on top of it.

If this piece of crust loses some of its material due to erosion, then it becomes lighter and "floats" higher in the mantle, where the erosional material is deposited and weighs down the crust, causing that area to become heavier and sink lower. This theory helps to explain the sinking of the geosynclines and the building of mountains.

Process Relationships

The close correlation among zones of earthquake and volcanic activity, geosynclines, mountain building, and subsidence suggests that these processes of crustal change are related.

Skill Assessments

Base your answers to questions 1 through 6 on your knowledge of Earth Science, the Reference Tables, and the three seismograms shown below. The seismograms were recorded at earthquake recording stations *A*, *B*, and *C*. The letters *P* and *S* on each seismogram indicate the arrival times of the compressional (primary) and shear (secondary) seismic waves.

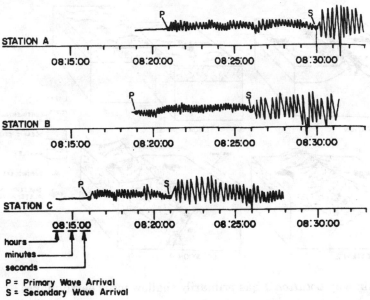

P = Primary Wave Arrival
S = Secondary Wave Arrival

1 Tell which station is farthest from the epicenter and explain how that can be determined.

2 How far is station *B* from the epicenter?

3 Explain how the speeds of P-and S-waves enable a seismologist to determine the distance to an epicenter.

4 The radius of each circle on the map at the right represents the distance from each seismographic station to the epicenter. Label the points representing the three recording stations (*A, B, C*) to correctly illustrate the position of the seismographic stations relative to the earthquake epicenter.

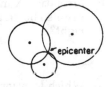

5 A fourth station recorded the same earthquake. The P-wave arrived, but the S-wave did not arrive. What is a possible explanation for the absence of the S-wave?

6 The epicenter distance from station *A* was calculated to be 7,600 kilometers. Approximately how long did the P-wave take to get to station *A*?

Base your answers to questions 7 through 10 on your knowledge of Earth Science, the Reference Tables, and the diagrams which represent geologic cross sections of the upper mantle and crust at four different Earth locations. In each diagram, the movement of the crustal sections (plates) is indicated by arrows and the locations of frequent earthquakes are indicated by symbols as shown in the key. Diagrams are not drawn to scale.

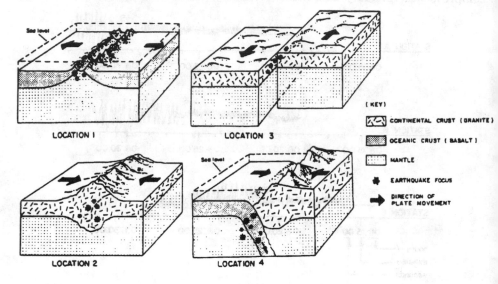

7   Explain why *Location 1* has primarily shallow focus earthquakes.

8   Which location would have some deep focus earthquakes? What is happening at this location that results in these deep focus earthquakes?

9   Which of the four locations most likely represents formation *X* in the profile shown at the right?

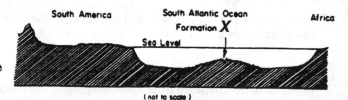

10   Which location might represent the formation of the Himalaya Mountains? Why?

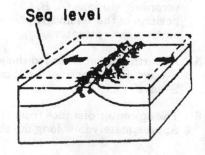

11   In the diagram at the right, draw arrows illustrating the motion of the convection currents in the mantle at this location.

Questions For Topic XII

1 The best evidence of crustal movement would be provided by
   1 dinosaur tracks found in the surface bedrock
   2 marine fossils found on a mountaintop
   3 weathered bedrock found at the bottom of a cliff
   4 ripple marks found in sandy sediment

2 The diagrams show cross sections of exposed bedrock. Which cross section shows the *least* evidence of crustal movement?

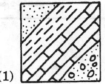 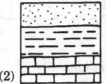 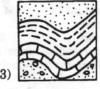 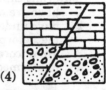

   (1)   (2)   (3)   (4)

3 The landscape shown in the diagram at the right is an area of frequent earthquakes.

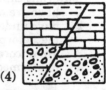

   This landscape provides evidence for
   1 converging convection cells within the rocks of the mantle
   2 movement and displacement of the rocks of the crust
   3 density differences in the rocks of the mantle
   4 differential erosion of hard and soft rocks of the crust

4 The epicenter of an earthquake is located near Massena, New York. According to the Reference Tables, the greatest difference in arrival times of the P- and S-waves for this earthquake would be recorded in
   1 Albany, New York   3 Plattsburgh, New York
   2 Utica, New York   4 Binghamton, New York

5 According to the Reference Tables, how long will a shear wave (S-wave) take to travel from an earthquake epicenter to a seismograph recording station 6,000 kilometers away?
   (1) 1 minute 40 seconds   (3) 9 minutes 20 seconds
   (2) 8 minutes   (4) 17 minutes

6 Through which zones of the Earth do compressional waves (P-waves) travel?
   1 crust and mantle, only
   2 mantle and outer core, only
   3 outer and inner core, only
   4 crust, mantle, outer, and inner core

7 At a seismic station, the arrival of an earthquake's P-wave is recorded three minutes earlier than the arrival of its S-wave. Approximately how far from the station is the earthquake's epicenter?
   (1) 700 km   (2) 1,400 km   (3) 1,900 km   (4) 5,600 km

8   The graph at the right
    shows the average velocities
    of P-waves traveling
    through various rock
    materials.

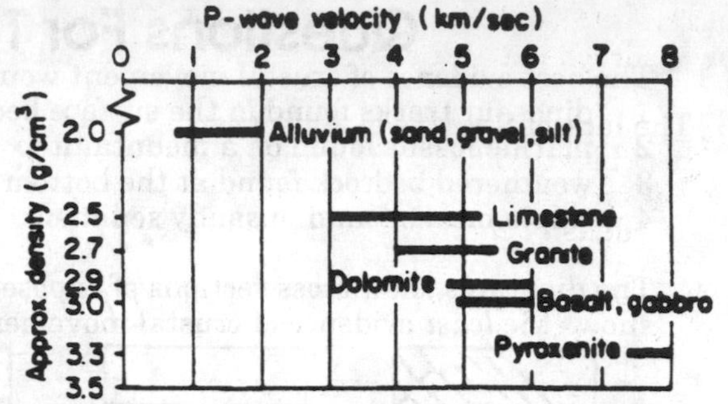

P-wave velocity (km/sec)

    The graph indicates that
    P-waves generally travel
    faster through rock
    materials that
    1   have greater density
    2   have undergone
        metamorphism
    3   are unconsolidated sediments
    4   are formed from marine-derived sediments

9   An earthquake occurred at 5:00:00 a.m. According to the Reference Tables,
    at what time would the P-wave reach a seismic station 3,000 km from the
    epicenter?
    (1) 5:01:40 a.m.   (3) 5:05:40 a.m.
    (2) 5:04:30 a.m.   (4) 5:10:15 a.m.

10 The circles on the map show the
    distances from three seismic stations,
    X, Y, and X, to the epicenter of an
    earthquake.

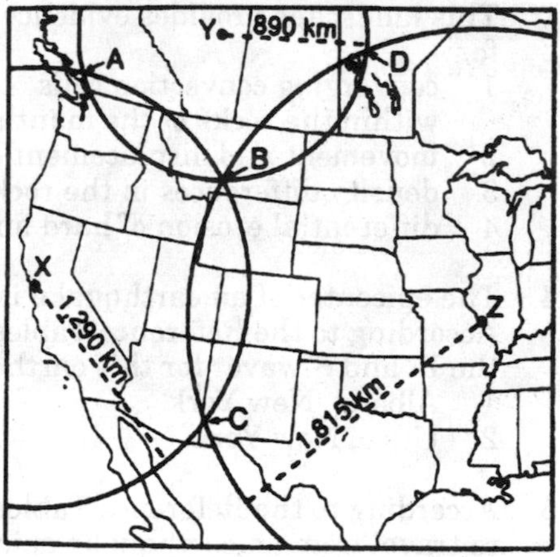

    Which location is closest to the
    earthquake epicenter?
    (1) A
    (2) B
    (3) C
    (4) D

11 An abrupt change in the speed of seismic waves is an indication that the
    1   seismic waves are colliding
    2   shear wave has overtaken the compressional wave
    3   waves are going into a material with different properties
    4   waves are passing through material of the same density

12 A seismographic station determines that its distance from the epicenter of
    an earthquake is 4,000 kilometers. According to the Reference Tables, if the
    P-wave arrived at the station at 10:15 a.m., the time of the earthquake's
    origin was
    (1) 10:02 a.m.   (3) 10:10 a.m.
    (2) 10:08 a.m.   (4) 10:22 a.m.

Base your answers to questions 13 and 14 on the diagram of the Earth showing the observed pattern of waves recorded after an earthquake.

13 The lack of S-waves in Zone 3 can best be explained by the presence within the Earth of
1 density changes
2 mantle convection cells
3 a liquid outer core
4 a solid inner core

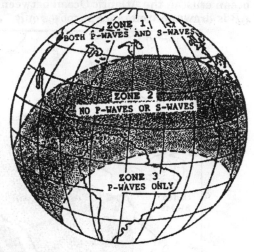

14 The location of the epicenter of the earthquake that produced the observed wave pattern most likely is in the
1 crust in Zone 1
2 mantle in Zone 2
3 crust in Zone 3
4 core of the Earth

Base your answers to questions 15 and 16 on your knowledge of Earth Science, the Reference Tables, and the map which shows three circles used to locate an earthquake epicenter. Five lettered locations, A, B, C, D, and E, are shown as reference points. Epicenter distances from three locations are represented by r_1, r_2, and r_3.

15 Location D is about 3,600 kilometers from the epicenter. What was the S-wave travel time to location D?
(1) 5 minutes 10 seconds
(2) 6 minutes 20 seconds
(3) 7 minutes 30 seconds
(4) 11 minutes 40 seconds

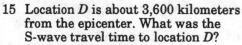

16 At which location could the seismogram below have been recorded?

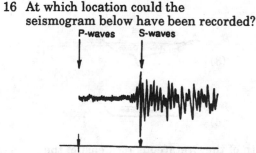

P-waves   S-waves

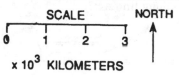

SCALE   NORTH

0   1   2   3

x 10³ KILOMETERS

10:12:00   10:15:40

(1) *A*   (2) *B*   (3) *C*   (4) *D*

Base you answers to questions 17-20 on your knowledge of Earth Science, the Reference Tables, and the map below. The map represents the age of the basaltic ocean crust in the Atlantic Ocean between the United States and Africa. Line *AB* is drawn for reference purposes only.

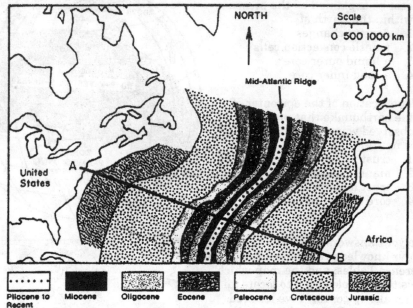

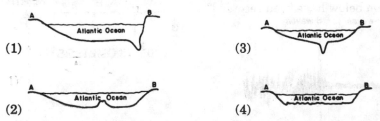

17 Which statement is best supported by the diagram?
1   The ocean crust is the same age along line *AB*.
2   The oldest ocean crust is located near the continents.
3   The age of the ocean crust increases from point *A* to point *B*.
4   Most of the ocean crust along line *AB* formed in the Paleozoic Era.

18 The age of formation of the ocean crust along line *AB* suggests that the United States and Africa are moving
1   eastward   3   closer together
2   westward   4   farther apart

19 Which diagram most closely represents the cross section of the ocean floor along line *AB*?

(1)

(3)

(2)

(4)

20 According to the diagram, the width of the Cretaceous rock east of the mid-Atlantic Ridge along line *AB* is approximately
(1) 1,000 km   (2) 1,200 km   (3) 1,600 km   (4) 4,000 km

Base your answers to questions 21-24 on your knowledge of Earth Science, the Reference Tables, and the diagrams. Diagram I is a map showing the location and bedrock age of some of the Hawaiian Islands. Diagram II is a cross section of an area of the Earth illustrating a stationary magma source and the process that could have formed the islands.

DIAGRAM I

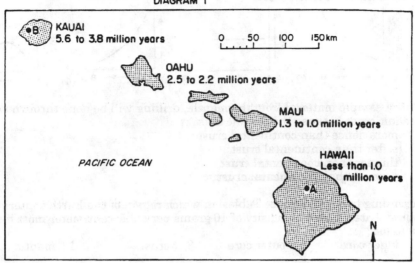

DIAGRAM II

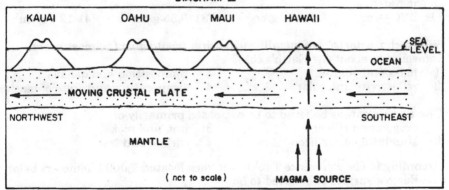

21 If each island formed as the crustal plate moved over the magma source in the mantle as shown in diagram II, where would the next volcanic island most likely form?
    1   northwest of Kauai   3   southeast of Hawaii
    2   northeast of Hawaii   4   between Hawaii and Maui

22 Volcanic activity like that which produced the Hawaiian Islands is usually correlated with
    1   nearness to the center of a large ocean
    2   sudden reversals in the Earth's magnetic field
    3   frequent major changes in climate
    4   frequent earthquake activity

23 Which of the Hawaiian Islands has the greatest probability of having a volcanic eruption?
1 Kauai   2 Oahu   3 Maui   4 Hawaii

24 Which graph best represents the ages of the Hawaiian Islands, comparing them from point A to point B?

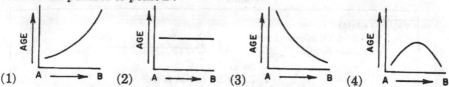

25 To get sample material from the mantle, drilling will be done through the oceanic crust because the oceanic crust is
1 more dense than continental crust
2 softer than continental crust
3 thinner than continental crust
4 younger than continental crust

26 According to the Reference Tables, in which region of the Earth's interior would material with a density of 10 grams per cubic centimeter most likely be found?
1 inner core   2 outer core   3 crust   4 mantle

27 According to the Reference Tables, what is the approximate average density of the Earth?
(1) 2.80 g/cm^3 (2) 5.52 g/cm^3 (3) 9.55 g/cm^3 (4) 12.0 g/cm^3

28 Which characteristic of metallic meteorites would give the most useful information about the Earth's core?
1 mass and volume   3 composition and density
2 reflectivity and color   4 temperature and state of matter

29 The Earth's core is believed to be composed primarily of
1 oxygen and silicon   3 iron and nickel
2 aluminum and silicon   4 carbon and iron

30 According to the Reference Tables, the rock located 1,000 kilometers below the Earth's surface is believed to be
1 liquid at approximately 1,800 °C   3 liquid at approximately 4,000°C
2 solid at approximately 3200°C   4 solid at approximately 4,500°C

31 According to the Reference Tables, approximately how far below the Earth's surface is the interface between the mantle and the outer core?
(1) 30 km   (2) 700 km   (3) 2,800 km   (4) 5,200 km

32 The primary cause of convection currents in the Earth's mantle is believed to be the
1 differences in densities of earth materials
2 subsidence of the crust
3 occurrence of earthquakes
4 rotation of the Earth

Earth's History — Interpreting Geologic History

Vocabulary To Be Understood In Topic XIII

Absolute Age (Dates)
Bedrock (Local Rock)
Carbon-14 Dating
Correlation
Extrusion, Intrusion
Fossil, Geologic Time Scale
Half-life
Index Fossil
Isotope (Radioactive)
Joint, Vein
Organic Evolution

Outcrop
Principle of Superposition
Radioactive Dating, Decay
Relative Age (Dates)
Rock Formation
Species
Unconformity
Uniformitarianism
Uranium-238
Volcanic Ash
Walking the Outcrop

A. Geologic Events

The history of the changing Earth is told in the Earth's **geologic events**, and the analysis, synthesis, and interpretation of these geologic events is a form of puzzle solving. Topics IX through XII reviewed most of the processes of Earth change. Topic XIII explores some of the techniques that geologists use to interpret geologic history recorded in the rocks.

Sequence Of Geologic Events

Knowing the sequence of the geologic events that took place during the formation of the Earth's crust makes it possible to develop a geologic history of the Earth and to better understand the forces that have and still are changing the Earth's crust.

Relative age is concerned with the sequence of geologic events that have occurred in an area as shown by the appearance of the rock layers. Relative age is not concerned with the actual ages of the rocks. This method uses sedimentary rock layers, igneous extrusions and intrusions, faults, folds, continuity, similarity of rock, fossil evidence, and volcanic time markers as clues to determine the probable order and conditions under which rock layers formed.

The actual age of a rock or fossil is called **absolute age**. The most accurate method of determining the absolute age of geologic events and rock is by techniques such as **radioactive dating**. Every radioactive element decays. A "parent" element emits radiation and particles until it is transformed by the loss into, eventually, a stable "daughter" element. Sometimes the "changing" element passes through a series of transformations into other radioactive elements before reaching this stability. Each radioactive element also has its own identifiable pattern and rate of decay.

How can the order in which geologic events occurred be determined?

Chronology Of Layers

The bottom layer in a series of horizontal sedimentary rock layers is the oldest, unless the series has been overturned or has had older rock thrust over it. This concept, called the **principle of superposition**, is used to determine the sequence in which a series of sedimentary layers was formed.

Principle Of Superposition

Igneous Intrusions And Extrusions

The rock layers through which igneous intrusions or extrusions cut are older than the intrusions or extrusions themselves, since the rock layers must be formed prior to the intrusion of magma or extrusion of lava.

Contact metamorphism of the rocks, through which the magma has moved, provides an additional clue to their relative age. Due to contact metamorphism, the rocks surrounding the intrusion are metamorphosed as a result of contact with hot magma.

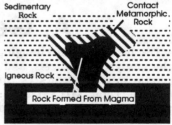

Contact Metamorphism

Faults, Joints, And Folds

Faults (cracks in the rock along which movement has occurred), joints (immovable cracks), and folds (bends in the rock strata) are younger than the rocks in which they appear. These distortions in rock occur due to changes in temperature and pressure.

Internal Characteristics

Fragments that occur within a rock are older than the rocks in which they are found, since previously they were formed from other rocks. However, cracks and **veins** (mineral deposits that have filled a rock crack or permeable zone) are younger than the rocks in which they occur.

Sedimentary rocks are younger than the sediments and the cements that formed them. Another kind of internal characteristic of rock layers is an **unconformity**, which is a zone where rocks of different ages meet. An unconformity is a "gap" in the geologic rock record due to erosion or nondeposition. It is usually seen as a buried erosional surface.

B. Correlation Techniques

The determination of the relative age of rock in geologic history can be accomplished through the use of correlation techniques. But, it is very important for the observer to distinguish actual *evidence* from possible *inferences*.

Correlation is the process of determining that the rock layers or geologic events in two separate areas are the same. Correlation involves observing the similarity and continuity of rock layers in different locations, comparing fossil evidence, and using volcanic time markers.

How can rocks and geologic events in one place be matched to another?

Continuity

When bedrock is exposed at the Earth's surface, it is called an **outcrop**, and correlation can be accomplished directly by **"walking the outcrop."** Over many years of destructional action, the landscape of a particular region may change greatly. For example, thick layers of level sedimentary rock could be cut into a wide valleys by the action of streams. After many years, the valleys will not resemble the original sedimentary formation. However, by careful examination of the rock strata (layers) exposed on opposite sides of the valleys, the geologist may be able to reconstruct the geologic history of the valleys.

Similarity Of Rocks

Rocks can often be tentatively matched on the basis of similarity in **appearance, color, and composition.** Referring back to the previous example, the geologist can use these rock characteristics to help figure out the puzzle of the valley's geologic history.

Fossil Evidence

The remains or traces of many once-living organisms, found almost exclusively in sedimentary rock, are called **fossils**. These fossils provide clues to the environment in which the organisms once lived. If a geologist finds fossils of a marine (or some other common) origin, evidence is provided that this sedimentary rock was formed in the sea. The geologist may then infer that this region was submerged at some time during geologic history.

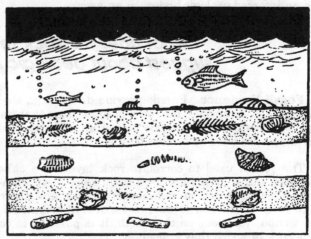

Fossil formation in the sea generally follows the same *Principle of Superposition* with the oldest fossils in the lower levels and younger above.

Volcanic Time Markers

A volcanic eruption is relatively short in duration when compared to the many years required for other constructional forces to build up the Earth's surface. When a volcano erupts, a layer of **volcanic ash** (fine particles of igneous rock ejected during the eruption) is rapidly deposited over a large area.

A layer of volcanic ash occurring between other layers of rock may serve as a time marker. Should a geologist discover a layer of volcanic ash buried between other layers of the sedimentary rock, and the actual date of the volcanic eruption is known, this time marker will provide very important information in determining the relative age of the rock layers above and below it.

Anomalies To Correlation

The process of "solving the puzzle" of geologic history appears fairly easy according to our discussion of rocks and geologic events and the correlation evidences. However, this *oversimplification* of the processes involved in determining geologic history may lead to *misconceptions*.

The geologist must exercise cautious interpretation to minimize this problem, since the very careful study of two similar rock formations may show that the rock formations are actually of different ages. Also, it is possible to find within a single formation, areas of different ages.

C. Determining Geologic Ages

The Rock Record

A close study of the rock record, using fossil evidence to develop a geologic time scale and erosional evidences to help fill in any gaps in the fossil record, can lead to an inferred geologic history of an area.

What does the rock record suggest about geologic history?

Fossil Evidence. Fossils provide direct (e.g., shells, bones) and indirect evidence (e.g., footprints, burrows) of organisms that had lived on Earth. Events in geologic history can often be placed in order according to relative age by using evidence provided by certain fossils.

The fossils used to correlate rock layers are called **index fossils** or **guide fossils**. Index fossils are used because of their wide-spread horizontal distribution (geographical) in sedimentary rocks and their relatively short period of existence on the Earth (narrow vertical distribution). By comparing these fossils in various locations on the Earth, it is possible to correlate the relative ages of the rock in which they appear.

Scale of Geologic Time. Geologists have subdivided geologic time into units, called **eons** (e.g., Phanerozoic, Proterozoic, Archean), **eras** (e.g., in the Phanero-

Geologic Time Scale

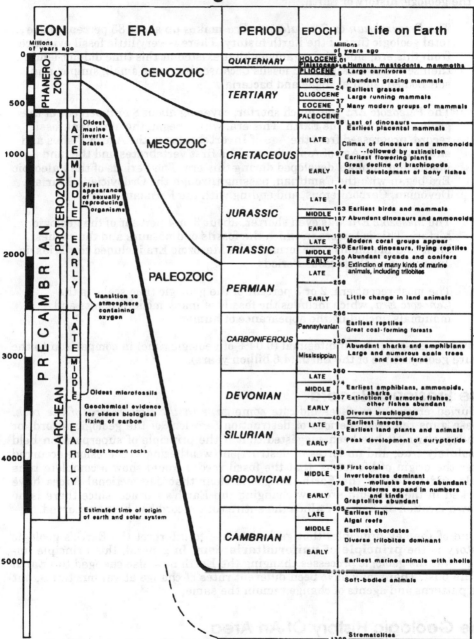

zoic Eon, Cenozoic, Mesozoic, and Paleozoic), **periods** (e.g., in the Mesozoic Era, Cretaceous, Jurassic, and Triassic), and **epochs**, based on the fossil evidence. However, note that most of the geologic past is devoid of a fossil record. (See the geologic time scale from the Reference Tables above.)

A general review of the **Geologic Time Scale** suggests the following sequence in the geologic history of Earth.

- The *Precambrian* or *Pre-Paleozoic Era* makes up about 85 percent of the total geologic time of the Earth history. There is very little fossil evidence from this era, since the organisms that existed at this time did not lend themselves to making good fossils because they were small, simple, and soft bodied (such as algae and bacteria).

- The *Paleozoic Era* was much shorter, covering about 8 or 9 percent of the geologic history of the Earth. This era, which began the abundant fossil record, progressed from the Age of Invertebrates to the Age of Fishes and ended with the Age of Amphibians. The first vertebrates and the land plants and animals developed during this era. The Periods of the Paleozoic Era began with the Cambrian, passing through the Ordovician, Silurian, Devonian, Carboniferous, and ending with the Permian.

- The *Mesozoic Era* was even shorter, about 3 or 4 percent of the geologic history. This is the era in which the fossils of dinosaurs and the earliest birds and mammals were formed. The Mesozoic Era included the Triassic, Jurassic, and Cretaceous Periods.

- The most recent era, 2 or 3 percent of the geologic time scale, is the *Cenozoic Era*, which includes the fossils of many modern plants and mammals, including the appearance of humans.

Human existence is infinitesimal (0.04% of geologic time) in comparison to the entire geologic time of the Earth (4.6 billion years).

The Erosional Record

Buried erosional surfaces indicate some *gaps in the time record of the rock*. These gaps represent periods of destruction (erosion) of the geologic record, or nondeposition. It has been suggested that if the principle of superposition held absolutely true, and no forces of destruction (weathering and erosion) occurred after the origin of life forms, that the fossil record would show a complete time scale and history of the Earth. But, it is clear that destructional forces have worked in the past and are now changing the Earth's surface, since there is no known location on Earth in which the entire rock record has been preserved.

One of the key principles that geologists use to interpret the Earth's geologic history is the **principle of uniformitarianism**. In general, this principle implies that the geologic processes changing the Earth now also changed the Earth in the past. There may have been different rates of change at various times, but the patterns and agents of change remain the same.

The Geologic History Of An Area

Using the evidence discovered in the rock record, the geologist can infer the geologic history of an area. The New York State geologic map and the geologic time scale may be used to illustrate the various portions of the rock record that have been preserved in New York State (see the Reference Tables).

Radioactive Decay

The evidences of the rock record discussed in this topic allow the geologist to make relative age estimates in the scale of geologic time and Earth history. To obtain absolute or actual age, the process of radioactive dating is used.

How can geologic ages be measured by using radioactive dating?

Decay Rates

When the spontaneous and natural nuclear breakdown of unstable atoms occurs, particles and energy are released. This constant, predictable process is called **radioactive decay**. By measuring the amount of radioactive isotope, compared to the amount of decay product (more stable form of the element), absolute age can be determined. The decay rate is unaffected by external factors, such as pressure and temperature, that would normally affect chemical reactions.

Some, but not all, rocks contain atoms whose nuclei undergo radioactive decay. This decay occurs as a random event and is not influenced by other changes occurring in the rock at the same time.

Radioactive Decay Data

Radioactive Element	Disintegration	Half-life
Carbon–14	$C^{14} \rightarrow N^{14}$	5.7×10^3 years
Potassium–40	$K^{40} \rightarrow Ar^{40}$	1.3×10^9 years
Uranium–238	$U^{238} \rightarrow Pb^{206}$	4.5×10^9 years
Rubidium–87	$Rb^{87} \rightarrow Sr^{87}$	4.9×10^{10} years

Half-lives Of Radioactive Substances

The half-life of a radioactive substance is the time taken for the activity of decay to reduce the total amount of radioactive substance in a material to half of its original amount. Therefore, by knowing the original content of a radioactive material and comparing it to the present content of the same radioactive material, the age of the material can be determined.

The half-lives of radioactive isotopes are different for different substances. Some radioactive isotopes, such as Carbon–14, have short half-lives and are good for dating recent organic remains (between 1,000 and 50,000 years).

Other radioactive substances, such as Uranium–238 which decays to stable Lead–206, have very long half-lives and are good for dating older rock formations (more than 10 million years). Uranium–238 has a half-life of about 4.5 billions years. The Earth itself is estimated to be 4.5 billion years old. Therefore, Uranium–238 that formed at the same time the Earth was formed has had time to undergo only one half-life.

Determining Age By Radioactive Decay

The age of a rock or fossil can often be inferred from the relative amounts of the undecayed (radioactive) substance and the decay product. For example, the actual age of a fossil can be inferred by the following method.

A sample piece of a fossil is taken and the amount of the Carbon–14 remaining in in that piece is measured at 0.5 grams. An equal sample of an existing organism, shows that the amount of original Carbon–14 was 2.0 grams. Since the radioactive decay of Carbon–14 is the loss of half of the total amount in the sample piece every 5,700 years, the sample was formed 11,400 years ago. Explanation: In the first 5,700 years of decay, the 2.0 grams were reduced to 1.0 gram; then, in the second 5,700 years of decay, the 1.0 gram was reduced in half again to 0.5 gram. Since one quarter of the original radioactive substance remained, the fossil had undergone two half-lives.

2 X 5,700 years (half-life of C¹⁴) = an age of 11,400 years

D. Fossil Record

Ancient Life. The study of the fossils found in sedimentary rock provides clues to the life that existed on the Earth in past eras and to the environments in which they lived.

What does the fossil record suggest about ancient life?

The Variety of Life Forms. Fossils give evidence that a great many kinds of animals and plants have lived in the past on the Earth in a great variety of environmental conditions. Most of these life forms do not exist (are extinct) on the Earth today. It is highly probable that in addition to the fossil types that have been found, there existed an even greater number of life forms that have left no traces (fossils) in the rock.

Evolutionary Development. A **species** is defined as organisms that are able to mate and produce offspring capable of continuing the same species. Within a species there are a great number of variations which can be observed, measured, and described.

Theories of **organic evolution** (how change occurs) have suggested that the variations within a species may provide some members of that species with a higher probability of survival. For example, a variation in the color of rabbits living in a snow covered Arctic region, such as white and a brown color difference, provide the white rabbit with a better chance to survive. The white rabbit may be able to better hide in the snow from predators. But, the brown rabbit would be easily seen. As a result, the white rabbit will have a higher probability of producing more offspring.

The similarity among some fossil forms of various time periods suggests a transition that may be a result of evolutionary development. Generally, the older rock formations contain more simple and marine life forms. The younger rock formations have the fossils of more complex land dwelling organisms.

NYS Fossils Through Geologic Time

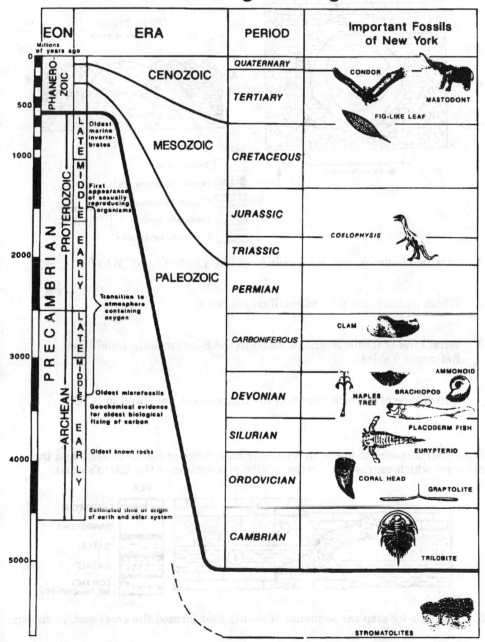

Skill Assessments

Base your answers to questions 1 through 4 on your knowledge of Earth Science, the Reference Tables, and the diagrams which represent the bedrock geology of an area outside New York State.

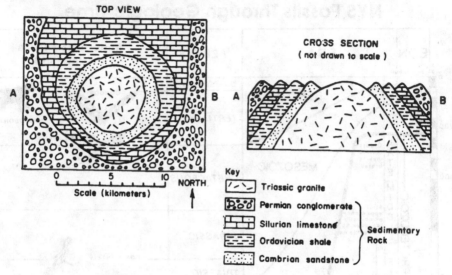

1 In which bedrock unit are fossils *least* likely to be found? Why?

2 Which bedrock could be 460 million years old?

3 What kind of fossils might be found in the Silurian limestone? [See Reference Tables]

4 Which type of bedrock shown in the diagrams is *not* found in NYS?

Base your answers to question 5 on your knowledge of Earth Science and the diagram which represents a cross section of a portion of the Earth's crust.

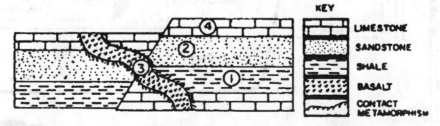

5 List step-by-step the sequence of events that formed the cross section shown.

Base your answers to questions 6 through 9 on your knowledge of Earth Science, the Reference Tables, and the chart which illustrates the geologic timespan and the relative abundance of species of ten types of animals and plants. The length of the timespan for each geologic time interval is not drawn to scale.

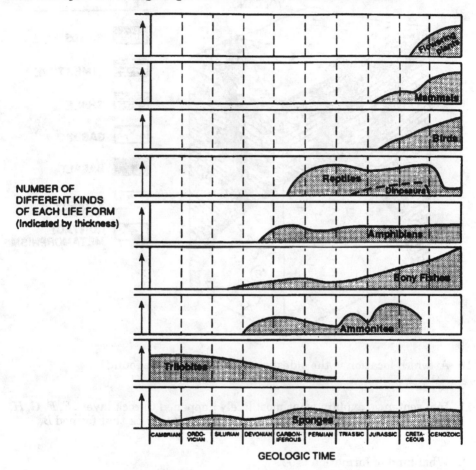

NUMBER OF
DIFFERENT KINDS
OF EACH LIFE FORM
(Indicated by thickness)

GEOLOGIC TIME

6   Which life form reached a peak three separate times in its existence on Earth?

7   When did the extinction of the dinosaurs occur?

8   Which life form appeared on Earth most recently?

9   Which life form existed for the greatest length of time?

Base your answers to questions 10 through 13 on your knowledge of Earth Science, the Reference Tables, and the diagram which represent a geologic cross section of a volcanic area where overturning of rock layers has *not* occurred.

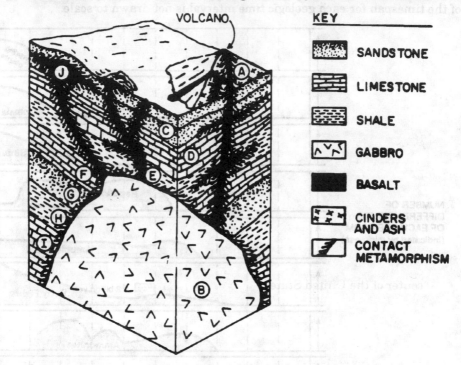

KEY

SANDSTONE

LIMESTONE

SHALE

GABBRO

BASALT

CINDERS AND ASH

CONTACT METAMORPHISM

10   At which location is the oldest rock most likely to be found?

11   In a sentence, explain what most likely happened to rock layers *E, F, G, H,* and *I* where they came in contact with the molten rock that formed *B*.

12   What kind of formation is *J*?

13   Which layer(s) might be sedimentary rock formed from marine-derived sediment?

Questions For Topic XIII

1   Volcanic ash layers may serve as excellent time markers in the geologic rock record because most volcanic ash
     1   contains fine-textured particles
     2   contains many minerals
     3   has a very low resistant to weathering
     4   is rapidly deposited over a wide geographic area

2　Unconformities (buried erosional surfaces) are good evidence that
　1　many life-forms have become extinct
　2　the earliest life-forms lived in the sea
　3　part of the geologic rock record is missing
　4　metamorphic rocks have formed from sedimentary rocks

3　Why are radioactive materials useful for measuring geologic time?
　1　The disintegration of radioactive materials occurs at a predictable rate.
　2　The half-lives of most radioactive materials are less than five minutes.
　3　The ratio of decay products to undecayed material remains constant in sedimentary rocks.
　4　Measurable samples of radioactive materials are easily collected from most rock types.

4　The map shows the relative age of the bedrock in the continental United States. The general age of the bedrock is found to be progressively

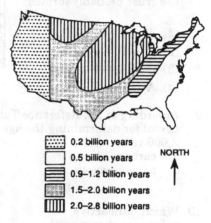

　1　older as an observer moves from the east and west coasts toward the center of the United States
　2　younger as an observer moves from the east and west coasts toward the center of the United States
　3　older as an observer moves across the United States from the east coast to the west coast
　4　younger as an observer moves across the United States from the east coast to the west coast

▨	0.2 billion years
☐	0.5 billion years
▤	0.9–1.2 billion years
▦	1.5–2.0 billion years
▥	2.0–2.8 billion years

NORTH ↑

5　In the Earth's geologic past there were long warm periods which were much warmer than the present climate. What is the primary evidence that these long warm periods existed?
　1　United States National Weather Service records
　2　polar magnetic directions preserved in the rock record
　3　radioactive decay rates
　4　plant and animal fossils

6　The diagram represents a sample of a sedimentary rock viewed under a microscope.

Which part formed first?

　1　the crack
　2　the pebbles
　3　the mineral vein
　4　the mineral cement

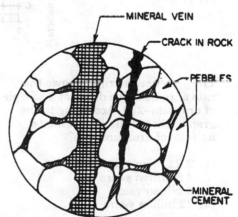

MINERAL VEIN

CRACK IN ROCK

PEBBLES

MINERAL CEMENT

7   An igneous intrusion is 50 million years old. What is the most probable age
    of the rock immediately surrounding the intrusion?
    (1) 10 million years   (3) 40 million years
    (2) 25 million years   (4) 60 million years

8   The diagrams show
    geologic cross sections of
    the same part of the
    Earth's crust at different
    times in the geologic past.

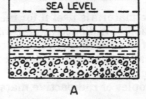

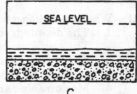

    Which sequence shows the
    order in which this part of
    the crust probably formed?

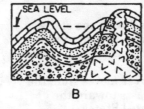

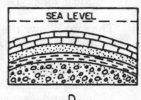

    (1) A → B → C → D
    (2) C → D → A → B
    (3) C → A → D → B
    (4) A → C → B → D

9   According to the Reference Tables, which radioactive element would be most
    useful for determining the age of clothing that is thought to have been worn
    2,000 years age?
    1   carbon–14   3   uranium–238
    2   potassium–40   4   rubidium–87

10 Which radioactive
    substance shown on the
    graph at the right has
    the longest half-life?

    (1) *A*
    (2) *B*
    (3) *C*
    (4) *D*

11 The graph at the right shows the
    rate of radioactive decay of a sample
    of uranium–238. According to the
    graph, what is the approximate
    half-life of uranium–238?

    (1) 2.2 billion years
    (2) 4.5 billion years
    (3) 6 billion years
    (4) 12 billion years

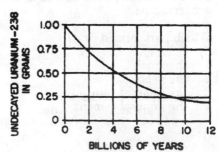

12　The diagram represents a clock used to time the half-life of a particular radioactive substance. The clock was started at 12:00. The shaded portion on the clock represents the number of hours one-half-life of this radioactive substance took to disintegrate.

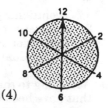

Which diagram best represents the clock at the end of the next half-life of this radioactive substance?

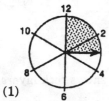　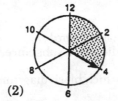　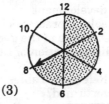　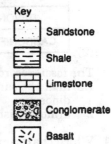

(1)　　　　　　(2)　　　　　　(3)　　　　　　(4)

13　A geologist uses carbon–14 to measure the age of some material found in a sedimentary deposit. If the half life of carbon–14 is 5.7×10^3 years and the sample shows that only $\frac{1}{4}$ of the original carbon–14 is left, the age of the sample is about
(1) 5,700 years　　　　　　(3) 17,100 years
(2) 11,400 years　　　　　　(4) 22,800 years

Base your answers to questions 14 and 15 on the geologic cross section diagram.

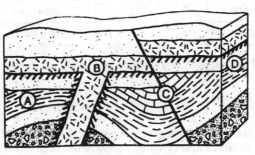

Key
Sandstone
Shale
Limestone
Conglomerate
Basalt

14　Which geologic event occurred most recently?
1　folding at *A*　　　　　　3　faulting at *C*
2　the intrusion at *B*　　　　4　the unconformity at *D*

15　The symbol ⟍⟍⟍⟍⟍ in the diagram most likely represents a

1　metamorphic rock in contact with an igneous rock
2　depression caused by underground erosion.
3　convection cell caused by unequal heating
4　large fault from an earthquake or crustal movement

16　From the study of fossils, what can be inferred about most species of plants and animals that have lived on the Earth?
1　They are still living today
2　They are unrelated to modern life forms.
3　They existed during the Cambrian Period
4　They have become extinct.

17 The Geologic Time Scale has been subdivided into a number of time units
    called periods based upon
    1   fossil evidence   3   rock types
    2   rock thickness   4   radioactive dating

18 Trilobite fossils from different time periods show small changes in
    appearance. These observations suggest that the changes may be the result
    of
    1   evolutionary development
    2   a variety of geologic processes
    3   periods of destruction of the geologic record
    4   the gradual disintegration of radioactive substances

19 According to the Reference Tables, which block diagram shows rock layers
    that have been overturned?

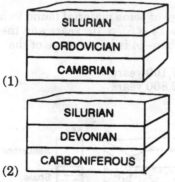

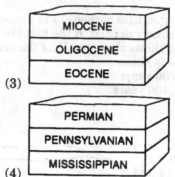

20 The fossil record provides evidence that primitive humans were alive on
    Earth at the same time as the
    1   dinosaurs   3   earliest birds
    2   armored fish   4   mammoths

21 Which characteristic of a fossil would make it useful as an index fossil in
    determining the relative age of widely separated rock layers?
    1   a wide time range and a narrow geographic range
    2   a wide time range and a wide geographic range
    3   a narrow time range and a wide geographic range
    4   a narrow time range and a narrow geographic range

22 Using the information in the Reference Tables, students plan to construct a
    geologic time line of the Earth's history from its origin to the present time.
    They will use a scale of 1 meter equals 1 billion years. What should be the
    total length of the students' time line?
    (1) 10.0 m   (2) 2.5 m   (3) 3.8 m   (4) 4.6 m

23 According to the Reference Tables, when did the armored fishes become
    extinct?
    1   before the appearance of dinosaurs
    2   before the appearance of land plants
    3   after the appearance of reptiles
    4   after the appearance of birds

Base your answers to questions 24 through 27 on your knowledge of Earth Science, the Reference Tables, and the diagram which represents a geologic cross section of part of the Earth's crust. Lines *XY* and *EF* represent erosional surfaces. The rock layers have not been overturned.

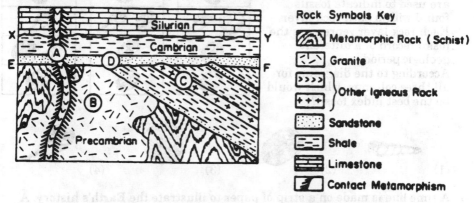

24　In which rock would a fossil eurypterid most likely be found?
1　Silurian limestone
2　Precambrian granite
3　Cambrian shale
4　metamorphic rock (schist)

25　What indicates that line *XY* is an erosional surface?
1　Igneous intrusion *A* cuts into the Silurian limestone.
2　Ordovician-age rock is missing between the Cambrian shale and the Silurian limestone.
3　Rock *C* has been partially removed by erosion.
4　Limestone cannot be formed directly on top of shale.

26　What is the most likely age of layer *D*?
1　Ordovician
2　Permian
3　Devonian
4　Cambrian

27　According to the "Generalized Bedrock Geology Map of New York State," where could Silurian limestone be found in the surface bedrock? [Refer to *Earth Science Reference Tables*.]
1　Old Forge
2　Watertown
3　Syracuse
4　Plattsburgh

28　A skull was discovered that has human characteristics and is about 2.8 million years old. Based on this information, during which epoch would early humans have existed? [Refer to the Reference Tables.]
1　Pliocene
2　Miocene
3　Oligocene
4　Eocene

29　Why are fossils rarely found in Precambrian rock layers?
1　Few Precambrian rock layers have been discovered.
2　Nearly all fossils from this era have been destroyed by glaciers.
3　Few rock layers were formed during the Precambrian Era.
4　Life that would produce fossils was not abundant during the Precambrian Era.

30 The geologic columns A, B, and
    C in the diagrams represent
    widely spaced outcrops of
    sedimentary rocks. Symbols
    are used to indicate fossils
    found within each rock layer.
    Each rock layer represents the
    fossil record of a different
    geologic period.
    According to the diagrams for
    all three columns, which would
    be the best index fossil?

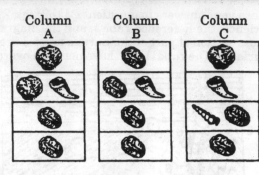

Column A   Column B   Column C

(1)   (2)   (3)   (4)

31 A time line is made on a strip of paper to illustrate the Earth's history. A
    length of 1.0 centimeter is used to represent 10. million years. According to
    the Reference Tables, what distance should be used to represent the length
    of the Mesozoic Era?
    (1) 0.179 cm   (2) 1.79 cm   (3) 17.9 cm   (4) 179 cm

32 Which two forms of life existed together on the Earth during the same time
    period?
    1   dinosaurs and mastodons   3   flowering plants and trilobites
    2   trilobites and birds   4   mastodons and flowering plants

33 Dinosaur footprints have been discovered in New York State surface
    bedrock. These footprints were most probably found in rocks formed during
    the
    1   Devonian Period   3   Precambrian Period
    2   Triassic Period   4   Permian Period

Landscape Development And Environmental Change

Vocabulary To Be Understood In Topic XIV

Dynamic Equilibrium
Escarpment
Faulting, Folding
Landscape
Landscape Region, Stage
Leveling (Destructional) Forces
Mountain

Plain
Plateau
Rock Resistance
Soil Association
Steam Drainage Pattern
Technology
Uplifting (Constructional) Forces

A. Landscape Characteristics

Landscapes are the results of the interaction of crustal materials, forces, climate, and man. This topic reviews the landscape characteristics that can be observed and measured, the relationships between landscape regions, and the development of landscapes.

Quantitative Observations

What are some landscape characteristics that can be observed and measured?

Hill Slopes. A hill slope includes the shape and gradient (slope) of land, characteristics of the land that have an affect on the patterns of stream drainage and soils. Hill slopes with distinctive shapes can be identified and measured with the use of "first hand" observations, like walking an outcrop, and the construction of models, like raised relief maps and contour maps.

Stream Patterns. Stream patterns can be identified which have measurable characteristics, such as the stream's gradient (slope), width, depth, drainage, and direction of flow.

Soil Associations. Although soils differ greatly in their composition, including mineral types, amounts of organic material (fertility), permeability rates, particle sizes and spaces, and the amount of soil horizon development, some soils have similar characteristics. These soils with similar characteristics are grouped together as a **soil association**. Soil associations may be found in the same area or in different locations, but they were developed from similar materials and in a similar manner.

Relationship Of Characteristics

How are landscape characteristics related?

Landscape Regions

Distinctive landscape regions can be identified by sets of landscape characteristics that seem to occur together. Obviously, the landscape characteristics of Long Island (lowlands) region are expected to be different from the landscape of the Adirondack (mountains) region.

Mountains – Landscape Regions

Landscapes may be classified as **mountains**, **plains** or **plateaus**. They may be distinguished from each other by their relief and internal rock structure. **Mountains** have high relief and deformed rock structure. **Plateaus** have moderate relief, high elevation, and horizontal rock structure. **Plains** are characterized by low relief and horizontal rock layers.

Generally, the boundaries between landscape regions are well defined. Landscape regions tend to be separated by mountains, large bodies of water, and other natural boundaries.

Since the continents are so large and usually bordering on oceans, they have several distinctive types of landscape regions. For example, the North American continent has regions that represent all of the major landscape regions of the Earth.

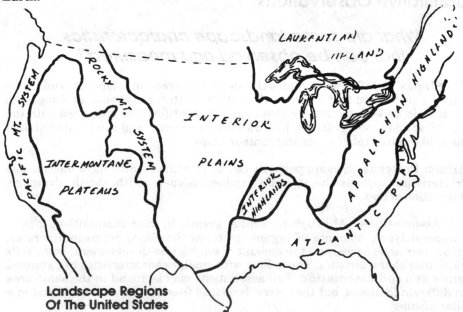

**Landscape Regions
Of The United States**

The continental United States has eight major landscape regions, including (west to east) the Pacific Mountain System, Intermontane Plateaus, Rocky Mountain, Interior Plains, Interior Highlands, Laurentain Upland, Appalachian Highlands, and Atlantic Plain.

The surface of New York State has many distinctive landscape regions because of a large number of bedrock variations. These include (north to south) Saint Lawrence Lowlands, Champlain Lowlands, Adirondack Highlands (mountains), Tug Hill Plateau, Erie-Ontario Lowlands (plains), Appalachian Uplands (plateau), Hudson-Mohawk Lowlands, New England Highlands, and the Atlantic Coastal Lowlands (plains).

B. Landscape Development

The processes of landscape development are very complex, producing various results in different areas. In general, most of the Earth changing processes are working all over the Earth, but not always with the same effects. For example, how wind and water affect the crust of the Earth is different over sedimentary and nonsedimentary regions. Some Earth materials, because of their characteristics are affected to a greater degree than other materials. Therefore, the generalizations discussed in this section, although generally true, may not be evident in all landscape regions.

Common Landscape Regions Of New York State

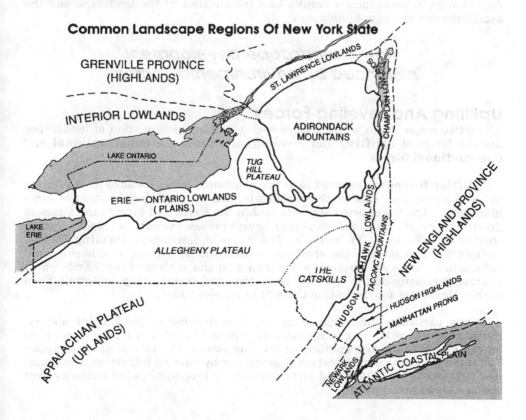

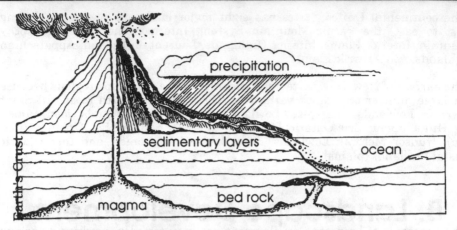

precipitation

sedimentary layers

ocean

Earth's Crust

bed rock

magma

Constructional And Destructional Forces

Environmental Factors

The environmental factors involved in landscape development, include uplifting and leveling forces (constructive and destructive, respectively), climate, bedrock characteristics, time, and human activities (positive and negative). *There is a delicate balance between the multiple environmental factors in all landscapes.* Any change in these factors results in a modification of the landscape and the establishment of a new equilibrium.

How is landscape development influenced by environmental factors?

Uplifting And Leveling Forces

The two major forces which oppose each other in the formation of landscapes are the forces of **uplifting** and **leveling**, also called the **constructional** and **destructional forces**.

Uplifting forces are referred to as constructional forces because their affect is to build mountains, enlarge the continents, and increase the elevation of some landscapes. The forces originate from within the Earth and include the forces of diastrophism (folding and faulting), earthquakes, volcanoes, and isostasy (reviewed in Topics X, XI, and XII). The forces of destruction, **leveling forces**, include the processes of the erosional – depositional system, weathering, and subsidence. These are forces found working on the surface of the Earth which operate primarily due to gravity, removing, and transporting materials from higher elevations and depositing them at lower elevations.

In any particular landscape or at any specific time in the Earth's history, either the force of leveling or uplifting may be dominant in a region, depending on the rate at which the uplifting or leveling occurs. The rate of uplift or subsidence may result in a modification of landscape by altering hill slopes, the drainage patterns of mountain and river regions, or orographic wind patterns over various landscape regions.

During the building of the Appalachian Mountains in the eastern United States, the force of uplifting was dominant. However, today the forces of leveling are dominant in that region since the elevation is decreasing. This change is due to weathering and erosion which are occurring faster than uplifting. In general, the uplifting forces are dominant where landscapes are increasing in elevation, and the forces of leveling are dominant when landscapes are becoming lower.

Climate Affect On Landscape Development

The climatic factors of temperature and moisture greatly affect the rate of the change in characteristics of landscapes. A change in climate may result in a modification of the landscape. For example, some scientists are concerned over the effects caused by a build up of carbon dioxide in the atmosphere. They suggest that this increase may bring about an increase in the Earth's average temperatures, resulting in a general warming of the Earth's surface. This could cause the melting of the polar ice regions, raising the sea level, thereby flooding low lying continental regions. A change in moisture from arid to humid would increase the rate of weathering and erosion and cause an angular landscape to become more rounded.

Landscape development can also affect climate. For example, the growth of a new mountain range could cause a humid region to become arid or a subsidence could change dry land into wetland (see Topic VII and VIII).

The steepness of hill slopes in an area is affected by the balance between weathering and the removal of materials. This can be observed in the hill slopes of arid and humid regions. Since arid regions have less water erosion, the hill slopes are generally more steep and angular, showing fewer signs of erosion. But in humid regions, there is more weathering, erosion, and deposition of materials causing the hills to become rounded and the hill slopes to be less steep. The Great Ice Age is an example of an extreme climatic change modifying the landscape.

Stream characteristics are affected by the climate. If the climate is dry (arid), the streams are intermittent (seasonal) and the water is usually collected in basins, remaining in the region. In wet (humid) climates, the streams are permanent, usually flowing into rivers and eventually into large bodies of water. Therefore, in arid regions most of the weathered sediment remains near the source, but in humid regions the sediment is often transported far from the source of the weathering.

Soil associations differ in composition depending on the climate. The lack of large amounts of water in arid regions causes a general decrease in the vegetation and animal populations, thereby decreasing organic materials in the soil and producing only thin or poorly developed soils. Arid soils are often high in mineral salts since they are not dissolved and carried away (leached) by water. The soil of a humid region is usually high in organic materials, low in mineral salts, less sandy, and better capable of supporting large populations of plants and animals.

Nonsedimentary Bedrock
Mountain Building
(non-existent leveling)

Bedrock
Varying Resistance
(escarpments)

Sedimentary Bedrock
(plateau)

Complex Bedrock
(faulting and
variations of leveling)

Samples Of Bed Rock Resistance

Bedrock Affect On Landscape Development

The rate at which landscape development occurs may be influenced by the bedrock of the region. Different landscape regions have various kinds of bedrock. **Rock resistance** is the ability of different rock types to resist the forces of weathering and erosion. For example, where hard bedrock is exposed to the environment, only a small amount of weathering and erosion occur. This is because the hard bedrock is more resistant to the forces of weathering and erosion than softer kinds of bedrock.

Where bedrock with varying degrees of rock resistance occur together, the softer bed rocks are weathered and eroded at a faster rate than the more resistant bedrock. The landscape surface is less regular than when there is a rock with resistance to weathering.

The shape and steepness of hills are affected by the local bedrock composition. Competent rocks (resistant to weathering and erosion) are responsible for the development of plateaus (high, stable landscapes with little or no distortion), mountains, and escarpments. These areas often have slow rates of change.

Weak or incompetent rocks (poor resistance to weathering and erosion) usually underlie valleys and other low level areas. Meandering streams, thick soil horizons, and deep layering usually characterize weak rock areas.

The difference between competent and weak areas can be observed in an **escarpment**, which is a steep slope separating two gently sloping surfaces. Many escarpments are the result of weathering forces on slopes with rocks of different resistance.

Structural features in bedrock, such as faults, folds, and joints, frequently affect the development of hill slopes. Generally, the more distortion in the rock mass of

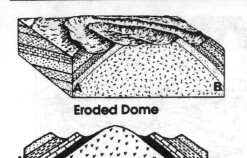

Eroded Dome

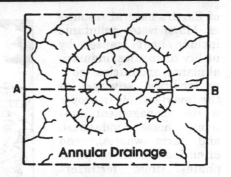

Annular Drainage

Stream Pattern On Bedrock With Varying Resistance

the hill slope, the greater the variety of changes occurring in the surface. The weathering effect is varied, since faults, folds, and joints expose rocks with different resistance to the forces of change.

Stream characteristics, including gradient (slope), width and depth, and drainage pattern and direction of flow, are controlled by bedrock characteristics. When the resistance of the bedrock over which the stream flows is consistent, a random drainage pattern forms, called **dendritic drainage**. **Trellis drainage** forms in regions of parallel folds and/or faults, whereas **radial drainage** is associated with volcanic cones and young domes. Annular (ring-shaped) drainage develops on domes as they become eroded.

Different bedrock composition produces different soil composition. Although climate is the most important factor in the determination of soil type, soils formed from parent material of different minerals will have different chemical compositions.

Time Affect On Landscape Development
The stage of development of a landscape is determined by the duration of time during which environmental factors have been active. Older landscapes generally show more effects of weathering and erosion than do younger landscapes.

Affect On Landscape Development
The activities of humans have altered the landscapes in many areas. Humans have modified landscapes both in negative and positive ways.

Negative Affects By Humans On The Landscape
Both intentional and unintentional **technological oversights** have brought about unplanned consequences that have destroyed or reduced the quality of life in many landscape regions. In some cases, technological advances have produced waste products which humans do not know how to dispose of safely. In other cases, industrial wastes are disposed of carelessly and/or criminally.

1) **Water pollution** is one of the most serious injuries to the landscape, since the Earth requires the purifying capabilities of water to clean and recycle many

of Earth's natural resources. Major water pollutants include **heat** (caused by utility companies, industrial plants, and nuclear power facilities), **sewage** (as the overflow of improper waste management systems, home, industrial, and farm wastes), and **chemicals** such as phosphates (from fertilizers and detergents), heavy metals (such as mercury from several industrial processes), **PCB's** (wastes from manufacturing), and **oil spills** (from industrial accidents, well drilling rigs, and fuel tanker cleaning).

2) **Air pollution** comes from **cars, mass transportation vehicles, homes, industrial plants**, and natural catastrophic events, such as **forest fires** and **volcanic eruptions** (e.g., Mt. St. Helens).

The major air pollutants include **carbon monoxide** (from burning), **hydrocarbons** (from fossil fuel burning cars, utilities, manufacturing), and **particulates** (dust, ash, smog). **Acid Rain** has now become a very serious result of air pollution. It develops from **nitrogen oxides** and **sulfur dioxide**, from industrial wastes and motor vehicles, combining with water droplets in the atmosphere.

One of the most serious problems with various air pollutants, including acid rain, is that weather fronts cause the pollutants to be carried to other landscape regions polluting the water and soil. The effects are not just localized.

3) **Biocides**, including **pesticides** and **herbicides**, have been used without complete knowledge of the possible harmful effects to the landscape. The environmental impact of various biocides has temporarily and in some cases permanently, contaminated the atmosphere, ground and surface water supplies, and the soil.

4) **Disposal problems** for the many toxic and even relatively harmless wastes of people's affluent life-style have become more and more serious in recent years. The discovery of hundreds of **toxic waste dumps**, such as the Love Canal in western New York State, have caused people and government to become concerned. Major disposal wastes often go into **landfills** and, when improperly disposed of, some wastes leak out of the dumps and into the ground water where they are transported to many different landscape regions.

Some of the disposal wastes are **solid** (such as cans, bottles, plastics, discarded appliances and cars), **chemical** (from chemical pesticides, industrial wastes and excesses, breakdown products from "thought to be" harmless wastes), and **nuclear** (radioactive particles, some remaining dangerous for millions of years).

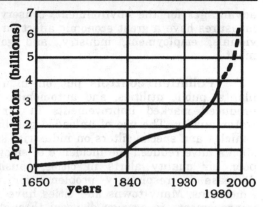

The human population growth has risen rapidly over the years, in part due to the ability to control diseases and modify environments to produce greater amounts of food and provide adequate shelter.

Overpopulation is considered by many scientists as the "ultimate threat" to the future stability of the environment.

Humans have been able to adjust the environment in ways that have made relatively uninhabitable land, habitable. However, this continued unchecked growth in population (at an exponential rate) has far exceeded the food and shelter–producing capacities of many world ecosystems.

Agricultural lands, water sources, grazing grasslands for livestock, and forests have been so badly misused as to have resulted in the starvation and extinction of total populations of plants and animals.

High population density areas are the most affected regions because of the concentration of the landscape pollution or the misuse of the landscape.

Positive Affects By Humans On The Landscape

Positive affects of human activities on landscapes have come through an increased awareness of various ecological and landscape interactions. Humans have begun to intervene in the widespread destruction, through the efforts of individuals, community groups, and conservation clubs. There are attempts to clean up toxic waste dumps, reclaim wasted lands, and prevent continued disruption of the environment.

1) **Population control** methods have been developed to balance the rate of population growth with the environment's capabilities for food production and shelter. Most modern countries have developed laws, produced guidelines for family planning, and provided education for their populations. However, many "third world" and "developing" countries still have not been able to solve their increasing population growth rates.

2) **The conservation of natural resources**, such as reforestation efforts and cover-cropping techniques, have helped to reclaim lost ecosystems. Water conservation practices have led to the use of land that was previously unusable for agricultural needs. States, including New York State, have passed laws requiring the recycling of cans and bottles, and encouraged the conservation of materials such as paper, plastics, and fuels for energy. In addition to the obvious

advantages for the environment, conservation measures have a great economic affect on individuals, employment, industry, and government.

3) **Pollution controls** put on industrial plants, public utilities, and automobiles have produced marked improvements in air and water quality. The use of unleaded gasoline in vehicles and special filters on industrial smoke stacks have reduced the incidence of smog and poor air quality alerts in cities, causing a decrease in respiratory problems for many Americans. Many towns and cities have made improvements in sewage disposal through the development and use of new sanitation techniques.

4) **Environmental Protection Laws** have been passed by state, local, and federal governments which regulate and guide the use of public lands and natural habitats. New York State designed a law (SEQR) which provides citizens with the opportunity to review and comment on the environmental impact of any proposed development, that may have a significant effect upon the landscape. Several state governments, including New York State, have passed freshwater and saltwater wetlands acts. These laws are designed to regulate and protect the large or unique wetland landscapes from development which would destroy them. These laws apply to both private and public owners and developers. These laws have helped to protect many species and maintain valuable landscapes for the future.

The future may be better if people and technological advances continue to keep the survival of our environment as a primary consideration, when making decisions that affect the landscape. It is possible to continue making a high standard of living and protecting all living things and the landscape, as long as there is a continued:

• awareness of the environment,

• a real concern for future generations, and

• the wise use of the Earth's energy and natural resources.

Skill Assessments

Base your answers to questions 1 through 4 on your knowledge of Earth Science, the Reference Tables, and the diagram which represents the geologic cross section and surface features of a portion of the Earth's crust. Locations *A, B, C,* and *D* are reference points of the Earth's surface.

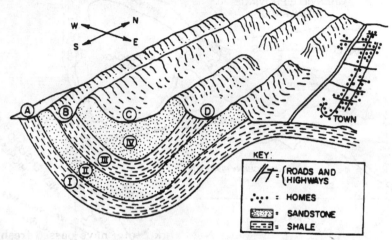

1   In a sentence or two, describe the type of rock structure underlying the regions located west of the town.

2   Index fossils are found in the surface bedrock under the town. Where else are fossils of the same type likely to be found?

3   What type of rock appears to be the most resistant to weathering?

4   Draw a diagram of the stream pattern that would most likely form in the region west of the town. What is the name of this stream pattern?

5   Both of the landscapes shown in the diagrams below are eroded plateaus. In a sentence or two, infer why their surfaces differ in appearance.

Base your answers to questions 6 through 11 on your knowledge of Earth Science, information in the Reference Tables, and the map below which shows the generalized landscape regions of New York State.

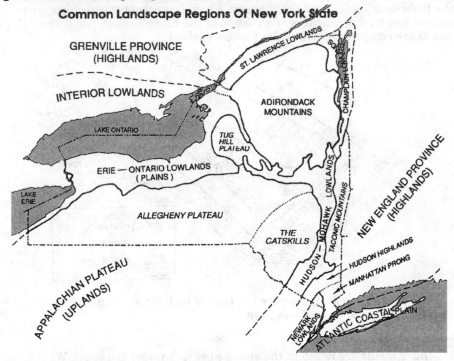

Common Landscape Regions Of New York State

6   Which river shown on the Generalized Bedrock Geology Map of New York State flows over bedrock with the greatest variation in ages? Where is the source of this river?

7   In a short paragraph, compare the rock age, rock type(s), and rock structure of the Adirondacks and the Catskills.

8   During what era was most of New York State's surface bedrock formed?

9   The state fossil of New York State is an *eurypterid*. In which landscape region would a geologist, looking for such fossils, have the best chance of finding them?

10   According to the Reference Tables, during which epoch was the advance and retreat of the last continental glacier? Make a list of glacial features currently found in New York State and indicate whether they are due to glacial erosion or glacial deposition.

11   Why does New York State have so many landscape regions?

Base your answers to questions 12 through 15 on your knowledge of Earth Science, information in the Reference Tables,and the stream patterns below which show the generalized landscape regions of New York State.

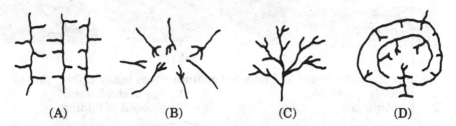

(A)   (B)   (C)   (D)

12 What is the pattern of stream *A*? On what what type of land surface would pattern *A* likely form?

13 What is the pattern of stream *B*? On what what type of land surface would pattern *B* likely form?

14 What is the pattern of stream *C*? On what what type of land surface would pattern *C* likely form?

15 What is the pattern of stream *D*? On what what type of land surface would pattern *D* likely form?

Questions For Topic XIV

1   One characteristic used to classify landscape regions as plains, plateaus, or mountains is
    1   type of soil   3   weathering rate
    2   amount of stream discharge   4   underlying bedrock structure

2   The diagrams show the same region of the Earth's crust at two different times.

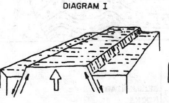

DIAGRAM I

DIAGRAM II
(Millions of years later)

These diagrams seem to indicate that landscape features are the result of
    1   only uplifting forces within the Earth's crust
    2   only leveling forces within the Earth's crust
    3   both uplifting and leveling forces acting on the Earth's crust
    4   neither uplifting nor leveling forces acting on the Earth's crust

3   Which characteristic of a landscape is usually the most difficult to observe?
    1   gradient of a stream   3   profile of the land
    2   type of soil   4   rate of erosion

4   Which cross-sectional diagram best represents a landscape region that resulted from faulting?

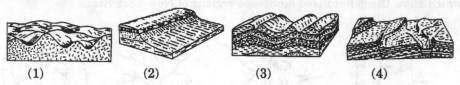

(1)   (2)   (3)   (4)

5   An area of gentle slopes and rounded mountaintops is most likely due to
1   climatic conditions   3   the age of the bedrock
2   earthquakes   4   the amount of folding

6   The diagram represents a section of the Earth's crust

This surface landscape was most likely caused by
1   folding of the crust
2   sinking of rock layers
3   erosion by valley glaciers
4   deposition of stream sediments

Base your answers to questions 7 through 11 on your knowledge of Earth Science and the diagram which represents a geologic cross section in which no overturning has occurred. The letters identify regions in the cross section.

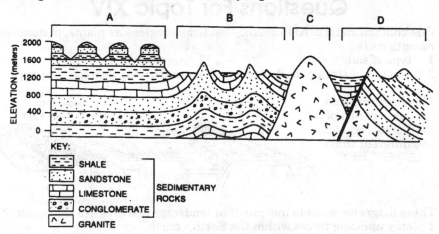

KEY:

SHALE
SANDSTONE
LIMESTONE   SEDIMENTARY ROCKS
CONGLOMERATE
GRANITE

7   The surface features in region *A* were produced primarily as a result of the process of
1   folding   2   faulting   3   erosion   4   glaciation

8   Which type of crustal movement is shown in region *B*?
1   faulting   3   jointing
2   volcanic eruption   4   folding

9 Which region shows the typical characteristics of plateau
(1) *A*   (2) *B*   (3) *C*   (4) *D*

10 Which region is least likely to have fossils in the surface bedrock?
(1) *A*   (2) *B*   (3) *C*   (4) *D*

11 Which stream drainage pattern would most likely be found on the surface is region *C*?

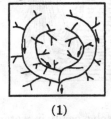   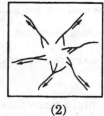      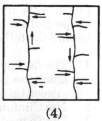

          (1)   (2)   (3)   (4)

12 Landscapes in which leveling forces are dominant over uplifting forces are often characterized by
  1   volcanoes   3   low elevations and gentle slopes
  2   mountain building   4   high elevations and steep slopes

Base your answers to questions 13 and 14 on the diagram that shows the surface landscape features and the internal rock structure of a cross section of the Earth's crust.

13 This landscape region would best be classified as
  1   an eroded plateau   3   folded mountains
  2   a coastal plain   4   volcanic mountains

14 How would this region most likely appear immediately after undergoing a period of glaciation?

   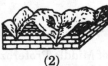      

          (1)   (2)   (3)   (4)

15 The well-defined boundaries of New York State's several distinct landscape regions are based on
  1   differences in bedrock composition and structure
  2   extreme differences in climate
  3   varieties of vegetation
  4   rate of sediment deposition

16 According to the Reference Tables, during which period of geologic history was the surface bedrock of the Catskills, NY, deposited?
  1   Cambrian   2   Pleistocene   3   Devonian   4   Triassic

17 The diagram shows a geologic cross section of the rock layers in the vicinity of Niagara Falls in western New York State.

LOCKPORT DOLOSTONE —
NIAGARA RIVER
ROCHESTER SHALE —
CLINTON LIMESTONE AND SHALE —
THOROLD SANDSTONE —
ALBION SANDSTONE AND SHALE —
WHIRLPOOL SANDSTONE —
QUEENSTON SHALE —

Which statement best explains the irregular shape of the rock face behind the falls?

1 The Lockport dolostone is an evaporite
2 The Clinton limestone and shale contain many fossils.
3 The Thorold sandstone and the whirlpool sandstone dissolve easily in water.
4 The Rochester and Queenston shale and the Albion sandstone and shale are less resistant to erosion than the other rock layers.

18 According to the Reference Tables, which event occurred most recently in New York State?
1 Taconian orogeny
2 extinction of the dinosaurs
3 formation of ancestral Adirondacks
4 intrusion of the Palisades Sill

19 Although the Adirondacks are classified as a mountain landscape, the Catskills are classified as a plateau landscape because of a major difference in their
1 amount of rainfall
2 bedrock structure
3 index fossils
4 glacial deposits

20 Salt deposits are often found in bedrock of Silurian age. Which landscape region in New York State most likely contains salt deposits?
1 Adirondack Mountains
2 Taconic Mountains
3 Atlantic Coastal Plain
4 Erie-Ontario Lowlands

21 According to the Reference Tables, the surface bedrock of the Tug Hill Plateau is composed primarily of
1 sedimentary rocks of Devonian age
2 sedimentary rocks of Ordovician age
3 igneous rocks of Cambrian age
4 intensely metamorphosed rocks of Middle Proterozoic age

22 The process of developing and implementing environmental conservation programs is most dependent on
1 the availability of the most advanced technology
2 the Earth's ability to restore itself
3 public awareness and cooperation
4 stricter environmental laws

23 If an area became more arid, the steepness of the slopes and sharpness of the landscape features would
1 decrease
2 increase
3 remain the same

Earth Science
Reference Tables

cm 1 2 3 4 5 6 7 8 9 10 11 12 13 14 15 16 17 18 19 20 21

Earth Science Reference Tables

Coelophysis
fossil footprints

1994 EDITION
The University of the State of New York
THE STATE EDUCATION DEPARTMENT
Albany, New York 12234

This edition of the Earth Science Reference Tables should be used in the classroom beginning in the 1993–94 school year. The first examination for which these tables will be used is the June 1994 Regents Examination in Earth Science.

Note: Centermeter Scale
may not be accurate. Use "fixed" ruler for measurements.

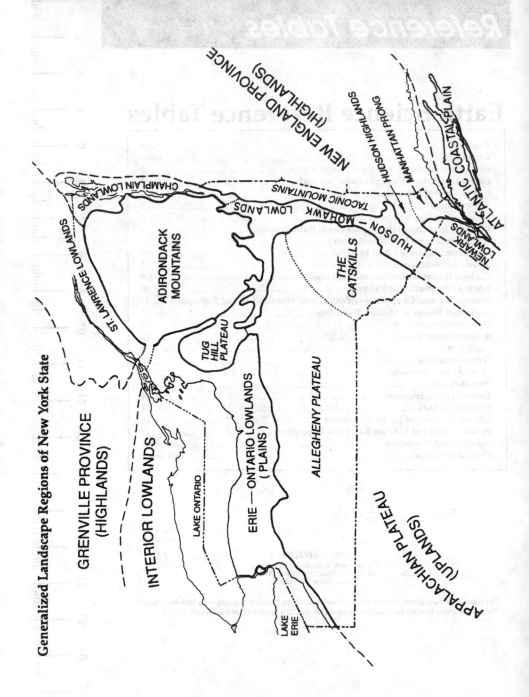

Generalized Landscape Regions of New York State

Generalized Bedrock Geology of New York State

COMPILED BY
GEOLOGICAL SURVEY
NEW YORK STATE MUSEUM
1989

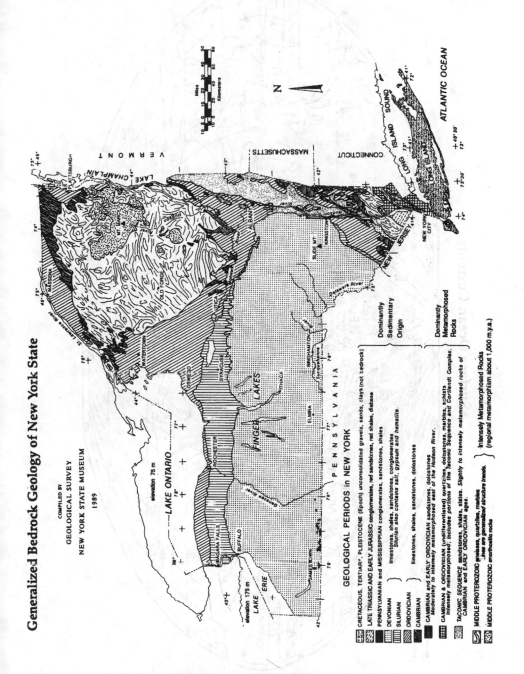

GEOLOGICAL PERIODS in NEW YORK

CRETACEOUS, TERTIARY, PLEISTOCENE (Epoch) unconsolidated gravels, sands, clays (not bedrock)

LATE TRIASSIC AND EARLY JURASSIC conglomerates, red sandstones, red shales, diabase

PENNSYLVANIAN and MISSISSIPPIAN conglomerates, sandstones, shales

DEVONIAN — limestones, shales, sandstones, conglomerates

SILURIAN — Silurian also contains salt, gypsum and hematite.

ORDOVICIAN — limestones, shales, sandstones, dolostones

CAMBRIAN — sandstones, shales, sandstones, dolostones

CAMBRIAN and EARLY ORDOVICIAN sandstones, dolostones
Moderately to intensely metamorphosed east of the Hudson River.

CAMBRIAN & ORDOVICIAN (undifferentiated) quartzites, dolostones, marbles, schists
Intensely metamorphosed.

TACONIC SEQUENCE sandstones, shales, slates. Slightly to intensely metamorphosed rocks of
CAMBRIAN and EARLY ORDOVICIAN ages. *Includes portions of the Taconic Sequence and Cortlandt Complex.
lines are generalized structure trends.*

MIDDLE PROTEROZOIC gneisses, quartzites, marbles

MIDDLE PROTEROZOIC anorthosite rocks
(regional metamorphism about 1,000 m.y.a.)

Dominantly Sedimentary Origin

Dominantly Metamorphosed Rocks

Intensely Metamorphosed Rocks

Surface Ocean Currents

→ WARM CURRENTS
→ COOL CURRENTS

Asia

Europe

Africa

Australia

Antarctica

North America

South America

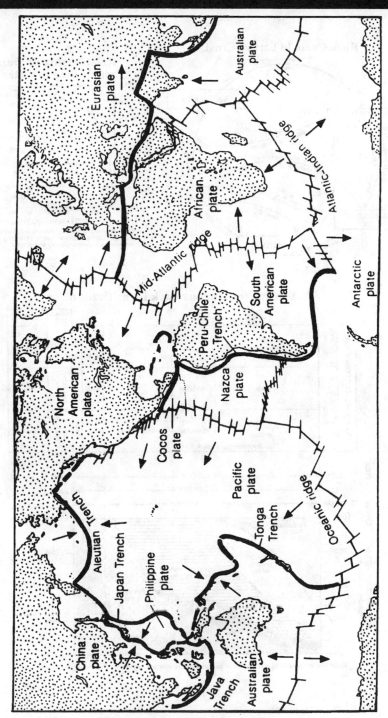

Tectonic Plates

Eurasian plate

Australian plate

Atlantic-Indian ridge

African plate

Mid-Atlantic ridge

South American plate

Antarctic plate

Peru-Chile Trench

North American plate

Nazca plate

Cocos plate

Pacific plate

Oceanic ridge

Aleutian Trench

Japan Trench

Tonga Trench

Philippine plate

China plate

Java Trench

Australian plate

Reference Tables Page 05

Rock Cycle in Earth's Crust

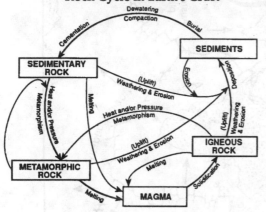

Relationship of Transported Particle Size to Water Velocity*

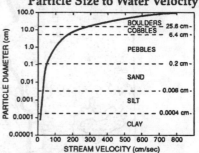

*This generalized graph shows the water velocity needed to maintain, but not start movement. Variations occur due to differences in particle density and shape.

Scheme for Igneous Rock Identification

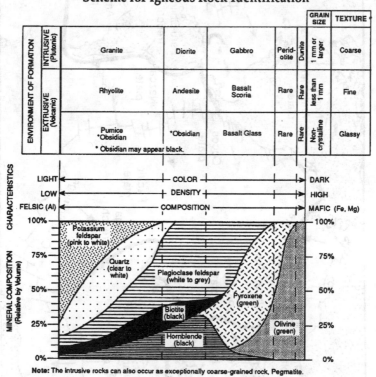

Note: The intrusive rocks can also occur as exceptionally coarse-grained rock, Pegmatite.

Reference Tables Page 06

Scheme for Sedimentary Rock Identification

INORGANIC LAND-DERIVED SEDIMENTARY ROCKS					
TEXTURE	GRAIN SIZE	COMPOSITION	COMMENTS	ROCK NAME	MAP SYMBOL
Clastic (fragmental)	Mixed, silt to boulders (larger than 0.001 cm)	Mostly quartz, feldspar, and clay minerals; May contain fragments of other rocks and minerals	Rounded fragments	Conglomerate	
			Angular fragments	Breccia	
	Sand (0.006 to 0.2 cm)		Fine to coarse	Sandstone	
	Silt (0.0004 to 0.006 cm)		Very fine grain	Siltstone	
	Clay (less than 0.0006 cm)		Compact; may split easily	Shale	

CHEMICALLY AND/OR ORGANICALLY FORMED SEDIMENTARY ROCKS					
TEXTURE	GRAIN SIZE	COMPOSITION	COMMENTS	ROCK NAME	MAP SYMBOL
Nonclastic	Coarse to fine	Calcite	Crystals from chemical precipitates and evaporites	Chemical Limestone	
	Varied	Halite		Rock Salt	
	Varied	Gypsum		Rock Gypsum	
	Varied	Dolomite		Dolostone	
	Microscopic to coarse	Calcite	Cemented shells, shell fragments, and skeletal remains	Fossil Limestone	
	Varied	Carbon	Black and nonporous	Bituminous Coal	

Scheme for Metamorphic Rock Identification

TEXTURE		GRAIN SIZE	COMPOSITION	TYPE OF METAMORPHISM	COMMENTS	ROCK NAME	MAP SYMBOL
FOLIATED	Slaty	Fine	CHLORITE MICA QUARTZ FELDSPAR AMPHIBOLE GARNET PYROXENE	Regional	Low-grade metamorphism of shale	Slate	
	Schistose	Medium to coarse			Medium-grade metamorphism; Mica crystals visible from metamorphism of feldspars and clay minerals	Schist	
	Gneissic	Coarse		(Heat and pressure increase with depth, folding, and faulting)	High-grade metamorphism; Mica has changed to feldspar	Gneiss	
NONFOLIATED		Fine	Carbonaceous		Metamorphism of plant remains and bituminous coal	Anthracite Coal	
		Coarse	Depends on conglomerate composition		Pebbles may be distorted or stretched; Often breaks through pebbles	Meta-conglomerate	
		Fine to coarse	Quartz	Thermal (including contact) or Regional	Metamorphism of sandstone	Quartzite	
			Calcite, Dolomite		Metamorphism of limestone or dolostone	Marble	
		Fine	Quartz, Plagioclase	Contact	Metamorphism of various rocks by contact with magma or lava	Hornfels	

Reference Tables Page 07

GEOLOGIC HISTORY OF NEW

EON	ERA	PERIOD	EPOCH	Life on Earth	Record in N.Y. Fossils			
					Root	Plants	Invertebrate	Vertebrate
PHANEROZOIC	CENOZOIC	QUATERNARY	HOLOCENE .01	Humans, mastodents, mammoths				
			PLEISTOCENE 2	Large carnivores				
		TERTIARY	MIOCENE	Abundant grazing mammals				
			OLIGOCENE 24	Earliest grasses / Large running mammals				
			EOCENE 37	Many modern groups of mammals				
			PALEOCENE 57					
			66	Last of dinosaurs / Earliest placental mammals				
	MESOZOIC	CRETACEOUS	LATE	Climax of dinosaurs and ammonoids --followed by extinction				
				Earliest flowering plants				
			EARLY	Great decline of brachiopods / Great development of bony fishes				
			144					
		JURASSIC	LATE 163	Earliest birds and mammals				
			MIDDLE 187	Abundant dinosaurs and ammonoids				
			EARLY 190					
		TRIASSIC	LATE 230	Modern coral groups appear / Earliest dinosaurs, flying reptiles				
			MIDDLE 240	Abundant cycads and conifers				
			EARLY 245	Extinction of many kinds of marine animals, including trilobites				
	PALEOZOIC	PERMIAN	LATE 258					
			EARLY	Little change in land animals				
			286					
		CARBONIFEROUS	Pennsylvanian	Earliest reptiles / Great coal-forming forests				
			320					
			Mississippian	Abundant sharks and amphibians / Large and numerous scale trees and seed ferns				
			360					
		DEVONIAN	LATE 374					
			MIDDLE 387	Earliest amphibians, ammonoids, sharks / Extinction of armored fishes, other fishes abundant				
			EARLY 408	Diverse brachiopods				
		SILURIAN	LATE 421	Earliest insects / Earliest land plants and animals				
			EARLY 438	Peak development of eurypterids				
		ORDOVICIAN	LATE 458	First corals				
			MIDDLE 478	Invertebrates dominant --mollusks become abundant				
			EARLY 505	Echinoderms expand in numbers and kinds / Graptolites abundant				
		CAMBRIAN	LATE	Earliest fish / Algal reefs				
			MIDDLE	Earliest chordates / Diverse trilobites dominant				
			EARLY	Earliest marine animals with shells				
			540	Soft-bodied animals				
PRECAMBRIAN	PROTEROZOIC							
	ARCHEAN		1300	Stromatolites				

Oldest marine invertebrates

First appearance of sexually reproducing organisms

Transition to atmosphere containing oxygen

Oldest microfossils

Geochemical evidence for oldest biological fixing of carbon

Oldest known rocks

Estimated time of origin of earth and solar system

Millions of years ago: 0, 500, 1000, 2000, 3000, 4000, 5000

YORK STATE AT A GLANCE

Important Fossils of New York	Tectonic Events Affecting Northeast North America	Important Geologic Events in New York	Inferred Position of Earth's Landmasses
CONDOR / MASTODONT / FIG-LIKE LEAF		Advance and retreat of last continental ice / Uplift of Adirondack region	**TERTIARY** 59 million years ago
		Sandstones and shales underlying Long Island and Staten Island deposited on margin of Atlantic Ocean / Development of passive continental margin / Kimberlite and lamprophere dikes	**CRETACEOUS** 119 million years ago
COELOPHYSIS	Rifting — Passive Margin	Atlantic Ocean continues to widen / Initial opening of Atlantic Ocean / Intrusion of Palisades Sill / Rifting	**TRIASSIC** 232 million years ago
CLAM		Massive erosion of Paleozoic rocks / Appalachian (Aleghanian) Orogeny caused by collision of North America and Africa along transform margin	**PENNSYLVANIAN** 306 million years ago
NAPLES TREE / AMMONOID / BRACHIOPOD / PLACODERM FISH / EURYPTERID / CORAL HEAD / GRAPTOLITE	Transform Collision	Catskill Delta forms / Erosion of Acadian Mountains / Acadian Orogeny caused by collision of North America and Avalon and closing of remaining part of Iapetus Ocean / Evaporite basins; salt and gypsum deposited / Erosion of Taconic Mountains; Queenston Delta forms / Taconian Orogeny caused by closing of western part of Iapetus Ocean and collision between North America and volcanic island arc	**DEVONIAN/MISSISSIPPIAN** 363 million years ago
TRILOBITE / STROMATOLITES	Rifting — Passive Margin — Subduction — Continental Collision	Iapetus passive margin forms / Rifting and initial opening of Iapetus Ocean / Erosion of Grenville Mountains / Grenville Orogeny: Ancestral Adirondack Mtns. and Hudson Highlands formed / Subduction and volcanism / Sedimentation, volcanism	**ORDOVICIAN** 458 million years ago

Reference Tables Page 09

Inferred Properties of Earth's Interior

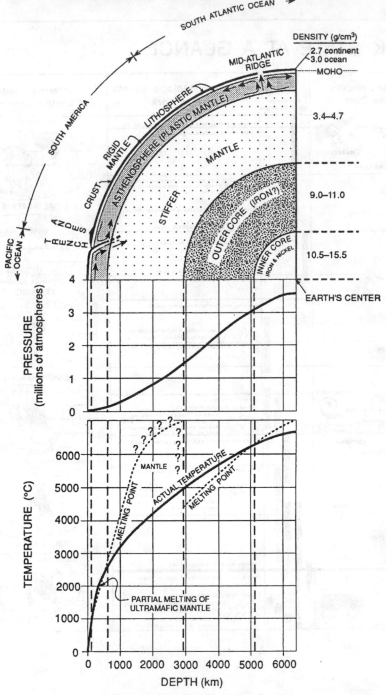

Reference Tables Page 10

Average Chemical Composition
of Earth's Crust, Hydrosphere, and Troposphere

ELEMENT (symbol)	CRUST		HYDROSPHERE	TROPOSPHERE
	Percent by Mass	Percent by Volume	Percent by Volume	Percent by Volume
Oxygen (O)	46.40	94.04	33	21
Silicon (Si)	28.15	0.88		
Aluminum (Al)	8.23	0.48		
Iron (Fe)	5.63	0.49		
Calcium (Ca)	4.15	1.18		
Sodium (Na)	2.36	1.11		
Magnesium (Mg)	2.33	0.33		
Potassium (K)	2.09	1.42		
Nitrogen (N)				78
Hydrogen (H)			66	

Earthquake P-wave and S-wave Travel Time

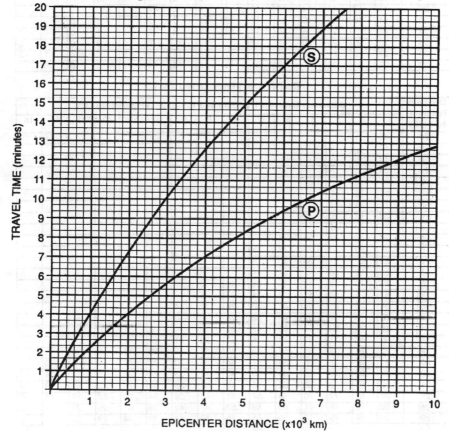

TRAVEL TIME (minutes)

EPICENTER DISTANCE (x10³ km)

Reference Tables Page 11

Dewpoint Temperatures

Dry-Bulb Temperature (°C)	Difference Between Wet-Bulb and Dry-Bulb Temperatures (C°)														
	1	2	3	4	5	6	7	8	9	10	11	12	13	14	15
−20	−33														
−18	−28														
−16	−24														
−14	−21	−36													
−12	−18	−28													
−10	−14	−22													
−8	−12	−18	−29												
−6	−10	−14	−22												
−4	−7	−12	−17	−29											
−2	−5	−8	−13	−20											
0	−3	−6	−9	−15	−24										
2	−1	−3	−6	−11	−17										
4	1	−1	−4	−7	−11	−19									
6	4	1	−1	−4	−7	−13	−21								
8	6	3	1	−2	−5	−9	−14								
10	8	6	4	1	−2	−5	−9	−14	−28						
12	10	8	6	4	1	−2	−5	−9	−16						
14	12	11	9	6	4	1	−2	−5	−10	−17					
16	14	13	11	9	7	4	1	−1	−6	−10	−17				
18	16	15	13	11	9	7	4	2	−2	−5	−10	−19			
20	19	17	15	14	12	10	7	4	2	−2	−5	−10	−19		
22	21	19	17	16	14	12	10	8	5	3	−1	−5	−10	−19	
24	23	21	20	18	16	14	12	10	8	6	2	−1	−5	−10	−18
26	25	23	22	20	18	17	15	13	11	9	6	3	0	−4	−9
28	27	25	24	22	21	19	17	16	14	11	9	7	4	1	−3
30	29	27	26	24	23	21	19	18	16	14	12	10	8	5	1

Relative Humidity (%)

Dry-Bulb Temperature (°C)	Difference Between Wet-Bulb and Dry-Bulb Temperatures (C°)														
	1	2	3	4	5	6	7	8	9	10	11	12	13	14	15
−20	28														
−18	40														
−16	48	0													
−14	55	11													
−12	61	23													
−10	66	33	0												
−8	71	41	13												
−6	73	48	20	0											
−4	77	54	32	11											
−2	79	58	37	20	1										
0	81	63	45	28	11										
2	83	67	51	36	20	6									
4	85	70	56	42	27	14									
6	86	72	59	46	35	22	10	0							
8	87	74	62	51	39	28	17	6							
10	88	76	65	54	43	33	24	13	4						
12	88	78	67	57	48	38	28	19	10	2					
14	89	79	69	60	50	41	33	25	16	8	1				
16	90	80	71	62	54	45	37	29	21	14	7	1			
18	91	81	72	64	56	48	40	33	26	19	12	6	0		
20	91	82	74	66	58	51	44	36	30	23	17	11	5	0	
22	92	83	75	68	60	53	46	40	33	27	21	15	10	4	0
24	92	84	76	69	62	55	49	42	36	30	25	20	14	9	4
26	92	85	77	70	64	57	51	45	39	34	28	23	18	13	9
28	93	86	78	71	65	59	53	47	42	36	31	26	21	17	12
30	93	86	79	72	66	61	55	49	44	39	34	29	25	20	16

Reference Tables Page 12

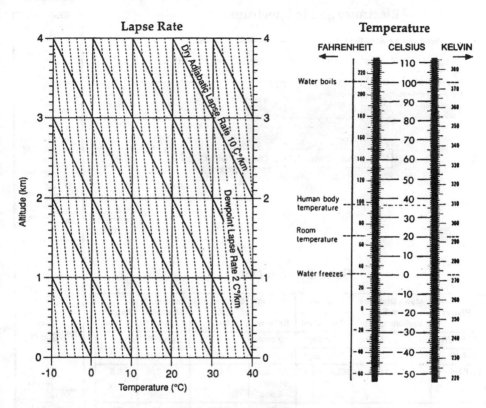

Lapse Rate

Temperature

Weather Map Information

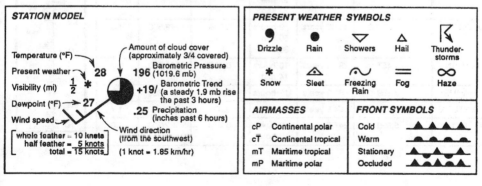

Reference Tables Page 13

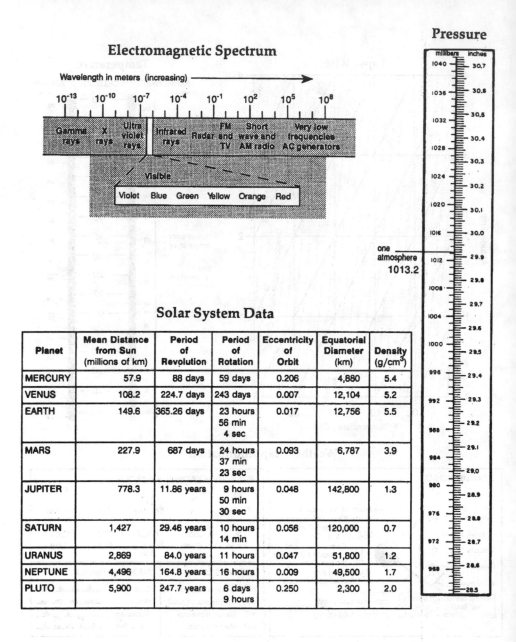

Electromagnetic Spectrum

Wavelength in meters (increasing) ⟶

10^{-13} 10^{-10} 10^{-7} 10^{-4} 10^{-1} 10^{2} 10^{5} 10^{8}

| Gamma rays | X rays | Ultra violet rays | Infrared rays | Radar and | FM and TV | Short wave and AM radio | Very low frequencies AC generators |

Visible

| Violet | Blue | Green | Yellow | Orange | Red |

Solar System Data

Planet	Mean Distance from Sun (millions of km)	Period of Revolution	Period of Rotation	Eccentricity of Orbit	Equatorial Diameter (km)	Density (g/cm³)
MERCURY	57.9	88 days	59 days	0.206	4,880	5.4
VENUS	108.2	224.7 days	243 days	0.007	12,104	5.2
EARTH	149.6	365.26 days	23 hours 56 min 4 sec	0.017	12,756	5.5
MARS	227.9	687 days	24 hours 37 min 23 sec	0.093	6,787	3.9
JUPITER	778.3	11.86 years	9 hours 50 min 30 sec	0.048	142,800	1.3
SATURN	1,427	29.46 years	10 hours 14 min	0.056	120,000	0.7
URANUS	2,869	84.0 years	11 hours	0.047	51,800	1.2
NEPTUNE	4,496	164.8 years	16 hours	0.009	49,500	1.7
PLUTO	5,900	247.7 years	6 days 9 hours	0.250	2,300	2.0

Pressure

millibars	inches
1040	30.7
1036	30.6
1032	30.5
1028	30.4
1024	30.3
1020	30.2
1016	30.1
	30.0
1012	29.9
1008	29.8
1004	29.7
1000	29.6
996	29.5
992	29.4
988	29.3
984	29.2
980	29.1
976	29.0
972	28.9
968	28.8
	28.7
	28.6
	28.5

one atmosphere **1013.2**

Selected Properties of Earth's Atmosphere

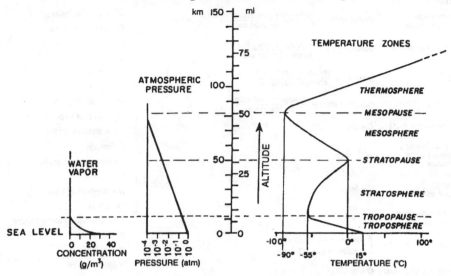

Planetary Wind and Moisture Belts in the Troposphere

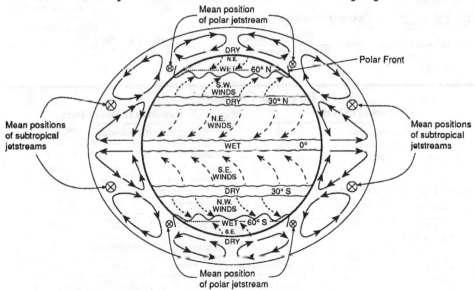

The drawing shows the locations of the belts near the time of an equinox. The locations shift somewhat with the changing latitude of the Sun's vertical ray. In the Northern Hemisphere the belts shift northward in summer and southward in winter.

Reference Tables Page 15

Equations and Proportions

Equations

Percent deviation from accepted value	deviation (%) = $\dfrac{\text{difference from accepted value}}{\text{accepted value}} \times 100$
Eccentricity of an ellipse	eccentricity = $\dfrac{\text{distance between foci}}{\text{length of major axis}}$
Gradient	gradient = $\dfrac{\text{change in field value}}{\text{change in distance}}$
Rate of change	rate of change = $\dfrac{\text{change in field value}}{\text{change in time}}$
Circumference of a circle	$C = 2\pi r$
Eratosthenes' method to determine Earth's circumference	$\dfrac{\angle a}{360°} = \dfrac{s}{C}$
Volume of a rectangular solid	$V = \ell wh$
Density of a substance	$D = \dfrac{m}{V}$
Latent heat	$\begin{cases} \text{solid} \longleftrightarrow \text{liquid} \quad Q = mH_f \\ \text{liquid} \longleftrightarrow \text{gas} \quad Q = mH_v \end{cases}$
Heat energy lost or gained	$Q = m\,\Delta T C_p$

C_p = specific heat
C = circumference
d = distance
D = density
F = force
h = height
H_f = heat of fusion
H_v = heat of vaporization
$\angle a$ = shadow angle
ℓ = length
s = distance on surface
m = mass
Q = amount of heat
r = radius
ΔT = change in temperature
V = volume
w = width
Note: $\pi \approx 3.14$

Proportions

Kepler's harmonic law of planetary motion	(period of revolution)^2 ∝ (mean radius of orbit)^3
Universal law of gravitation	force ∝ $\dfrac{\text{mass}_1 \times \text{mass}_2}{(\text{distance between their centers})^2}$ $\left(F \propto \dfrac{m_1\,m_2}{d^2} \right)$

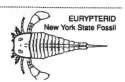

EURYPTERID
New York State Fossil

Physical Constants

Properties of Water

Latent heat of fusion (H_f)	80 cal/g
Latent heat of vaporization (H_v)	540 cal/g
Density (D) at 3.98°C	1.00 g/mL

Specific Heats of Common Materials

MATERIAL		SPECIFIC HEAT (C_p) (cal/g·C°)
Water	solid	0.5
	liquid	1.0
	gas	0.5
Dry air		0.24
Basalt		0.20
Granite		0.19
Iron		0.11
Copper		0.09
Lead		0.03

Radioactive Decay Data

RADIOACTIVE ISOTOPE	DISINTEGRATION	HALF-LIFE (years)
Carbon-14	$C^{14} \rightarrow N^{14}$	5.7×10^3
Potassium-40	$K^{40} \begin{smallmatrix} \nearrow Ar^{40} \\ \searrow Ca^{40} \end{smallmatrix}$	1.3×10^9
Uranium-238	$U^{238} \rightarrow Pb^{206}$	4.5×10^9
Rubidium-87	$Rb^{87} \rightarrow Sr^{87}$	4.9×10^{10}

Astronomy Measurements

MEASUREMENT	EARTH	SUN	MOON
Mass (m)	5.98×10^{24} kg	1.99×10^{30} kg	7.35×10^{22} kg
Radius (r)	6.37×10^3 km	6.96×10^5 km	1.74×10^3 km
Average density (D)	5.52 g/cm^3	1.42 g/cm^3	3.34 g/cm^3

Additional
Skill Assessments

Skill Assessments

Base your answers to questions 1 through 5 on your knowledge of Earth Science and the graph which shows the rate of rainfall during a storm and the stream discharge of a nearby stream on July 19.

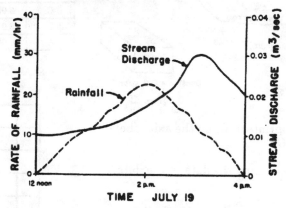

1   For how many hours did the rainstorm last?

2   What was the maximum rainfall rate? What was the maximum stream discharge?

3   In a sentence or two, explain why the time of maximum stream discharge occurred after the time of maximum rate of rainfall.

4   If the graph shown resulted from a storm in a rural area, predict how the graph might change if the same storm occurred in a nearby urban area.

5   Another rainstorm with the same characteristics as the rainstorm shown on the graph starts at 4 p.m. on July 19. Predict how maximum stream discharge from this second storm would compare with the maximum discharge from the first storm.

Base your answers to questions 6 through 13 on your knowledge of Earth Science, the Reference Tables, and the information given in the diagrams and data table. *Diagram I* represents a cross section of a lake. Letters A through E identify locations in the lake. The data table shows the depth and temperature of the water at each location. *Diagram II* represents a solid block of a nonporous uniform material.

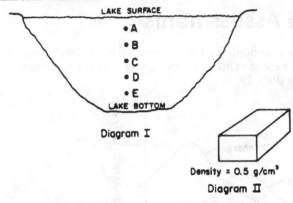

Diagram I

Density = 0.5 g/cm³

Diagram II

DATA TABLE

Location	Depth (m)	Temperature (°C)
A	2.00	15
B	4.00	10
C	6.00	7
D	8.00	5
E	10.00	4

6   Mark an appropriate scale on the axis labeled *Temperature °C*.

7   Plot a line graph using the data in the data table.

8   As depth increases, what happens to the temperature?

9   What is the relationship between the *rate* of temperature change and the depth?

10   Determine the temperature gradient in °C/m from A to E.

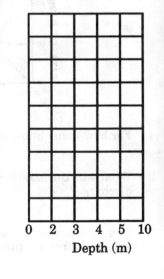

11   A student incorrectly measured the depth from the surface to D as 7.00 meters. What is the student's percent deviation?

12   The solid block of material shown in *Diagram II* is placed in the lake. Predict what will happen to it. In a sentence or two, explain your prediction.

13   If the solid block of material is compressed to one-fourth its original size and placed in the lake, predict what will happen to it. In a sentence or two explain your prediction.

Base your answers to questions 14 through 20 on your knowledge of Earth Science, the Reference Tables, and the map at the right which represents a section of a surface weather map. Letters A through G represent positions on the surface.

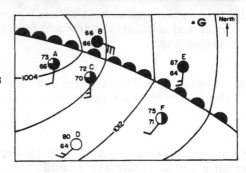

14 What kind of weather front is shown on the map?

15 In what compass direction is the front moving?

16 Estimate the barometric (air) pressure at station E and write it at the appropriate place on the station model at E.

17 Which station has the lowest barometric (air) pressure?

18 At which station does the reported data show the relative humidity to be 100%? Explain how you can tell.

19 Draw a cross-sectional diagram of the atmosphere between stations F and E showing the shape of the frontal surface, cloud cover, and use arrows to show the direction of air-mass movement.

20 Write a short weather forecast for station G for the next twenty-four hours. Include in your forecast any anticipated changes in temperature, air pressure, and sky conditions.

Base your answers to questions 21 through 25 on the Reference Tables, the topographic map, and your knowledge of Earth Science. The topographic map represents elevation contours measured in meters. Four straight lines, AB, CD, EF, and GH have been drawn for reference purposes.

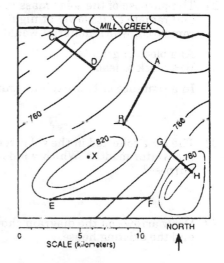

21 What is the contour interval of the map?

22 Label each contour line with the appropriate elevation placing the elevation number in the blank interruption shown in the contour line.

23 In what compass direction is Mill Creek flowing? Explain how you can tell.

24 Write the equation for determining gradient. Substitute the data from the map and calculate the gradient from *C* to *D* labeling your answer with the proper units.

25 If uplifting and leveling are in a state of dynamic equilibrium, what will happen to the average elevation of the area shown on this map?

Base your answers to questions 26 through 30 on your knowledge of Earth Science, the Reference Tables, and the diagrams. *Diagram I* shows the paths of the Sun in relation to a house in New York State on June 21 and on December 21. *Diagram II* shows the plan of a solar room which is to be added to the house in *Diagram I*.

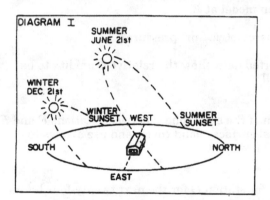

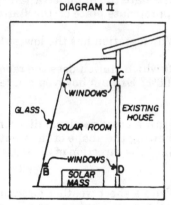

26 On which side of the house should the solar room be added to gain the greatest intensity of insolation?

27 The purpose of the solar mass in *Diagram II* is to absorb heat during the day and radiate heat slowly at night. If equal masses of the materials below are compared, which material would be the most effective solar mass?

A) a block of granite   C) dry air sealed in a black tank
B) a block of lead   D) water sealed in a black tank

In a sentence or two, explain your choice.

28 The heat collected in the solar room is used to heat the existing house during the winter months. Which windows should be opened to best serve this purpose?

29 Draw arrows on the diagram showing the air flow between the solar room and the existing house.

30 On *Diagram I* draw the position of the noon Sun on March 21 and use dashed lines to indicate the path of the Sun of the Sun on that date. Label the line "March 21st."

Base your answers to questions 31 through 35 on your knowledge of Earth Science and the diagrams. *Figure I* represents the physical setup of an energy absorption investigation. A black metal container and a shiny metal container are placed equal distances from a lamp heat source. A thermometer is inserted in each container. Inside air-temperature measurements are taken for 12 minutes while the lamp is on. *Figure II* is a graph of the temperatures recorded for the 12-minute period.

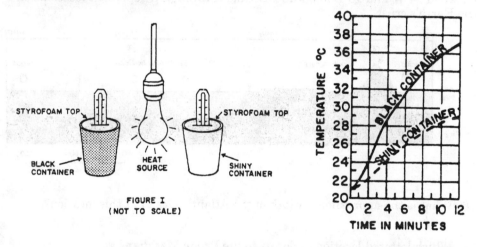

FIGURE I
(NOT TO SCALE)

FIGURE II

31 What is the primary process by which electromagnetic energy is being transferred from the heat source to the metal containers?

32 Compare the amount of heat received by each container during the 12 minutes.

33 At what rate, in °C/min., did the air temperature in the black container rise in the first 10 minutes?

34 In a sentence or two, account for the difference in the heating rates of the air in the black and shiny containers.

35 After 12 minutes the lamp was turned off and removed from the area around the containers allowing the containers to cool. Predict which can will cool faster. In a sentence or two, explain the reason for your prediction.

Base your answers to questions 36 through 40 on the Reference Tables, the diagram, and your knowledge of Earth Science. The diagram is a cross section of the major surface features of the Earth along the Tropic of Capricorn (23½°S) between 74°W and 15°E longitude. Letters A through G represent locations on the Earth's crust.

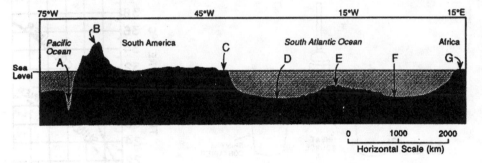

36 What is the approximate width of the Atlantic Ocean at this location?

37 Which lettered location is closest to the Prime Meridian?

38 Where is a mid-ocean ridge located?

39 A geologist took samples of rock at each location. Which two locations have bedrock of approximately the same age? Explain why.

40 On the diagram at the right draw arrows showing the probable direction of crustal plate movement and the direction of mantle convection currents under position A.

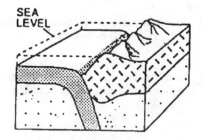

Base your answers to questions 41 through 44 on your knowledge of Earth Science, the Reference Tables, and the diagram. The diagram represents the Earth's tilt on its axis and shows the direction to the Sun and to the Moon on a particular day of the year. The shaded portion represents the nighttime side of the Earth.

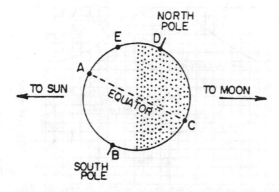

41 At which lettered position would an observer not be able to photograph the Moon at any time during the following 24 hours? In a sentence or two explain why.

42 Compare the number of daylight hours at location A with the number of daylight hours at location C.

43 At which position would the altitude of the North Star (Polaris) be closest to 90°?

44 On this day, as an observer travels from position D to position A, explain what happens to

a) the altitude of Polaris above the horizon

b) the duration of insolation

c) the angle of insolation

d) the distance of the observer from the center of the Earth.

Base your answers to questions 45 through 48 on the graph and your knowledge of Earth Science. The graph illustrates the relationship between average air temperature, average soil temperature at a depth of 1 meter, and the duration of insolation for a New York State location.

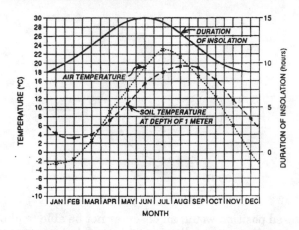

45 When the highest air temperature occurred what was happening to the duration of insolation?

46 In one or two sentences, explain why maximum air temperature occurred after the time of maximum duration of insolation.

47 What was the average rate of soil temperature change, in C°/month, for the period of time between mid-March and mid-July?

48 The air temperatures at this New York State location are lower in winter than in summer. In a sentence or two explain why.

Absolute Age (Date) (167, 173) – dates in Earth's history arrived at by dating radioactive rocks and measured without reference to any other event.

Absolute Humidity (84) – related to vapor pressure, the measurable weight of water vapor in a specific volume of air.

Absolute Zero (53) – lowest temperature theoretically possible; no electromagnetic energy radiated; 0 K (Kelvin) = –273° C. = –459.67° F.

Absorption (55, 69) – opposite of radiation; the concentration of a substance through (permeating) a surface (see adsorption).

Absorption of Solar Energy (69) – concentration of the electromagnetic radiation of the Sun through the Earth's atmosphere, including ultraviolet, infrared, visible wavelengths.

Acceleration (7) involves the mass of an object, a length (distance) of movement, and a unit of time of movement.

Acid Rain (101) – weak acid precipitation from industrial waste gases.

Actual Evapotranspiration (103) – see water budge; due to evaporation and transpiration, the actual amount of water lost from a given area over a specific amount of time, expressed in millimeters of water.

Adiabatic Changes (87) – changes that occur without any loss or gain of energy, such as in heat changes.

Aerobic Bacteria (102) – bacteria that use oxygen in energy production.

Aerosols (69, 79) – the mixtures of small particles suspended in a liquid or gas; such as fog, smog, muddy water, etc.

Air Mass (80) – large area of air within the lower atmosphere having generally the same temperature and humidity at any given level.

Air Pressure (78) – see weather.

Altitude (24) – a celestial object's angular distance above the horizon; the vertical distance of one point above the Earth's surface at sea level.

Anaerobic Bacteria (102) – bacteria that do not use oxygen in their production of energy, but usually carry on a form of fermentation.

Angle of Insolation (66) – angle that the rays of the Sun hit the surface of the Earth; decreases with an increase in latitude; increases as the angle approaches perpendicular; varies with the time of day and day of the year.

Angular Diameter (41) – See Apparent Planetary Diameter

Anticyclone (80) – HIGH, the opposite of a cyclone (LOW), having high pressure and clockwise winds in the northern hemisphere.

Aphelion (40, 45) – point on the Earth's orbit when the Earth is the farthest from the Sun (152 million kilometers); occurring on July 1st; point of Earth's slowest orbital speed.

Apparent Daily Motion (37, 40) – perceived movement of celestial objects as seen from the Earth, circular, constant, daily, and cyclic.

Apparent Planetary Diameter (41) – not actual measured diameter, but the perceived diameter of a planet, changing according to the distance from observer.

Apparent Solar Day (42, 43,) – time (24+ hours) required for the Sun to cross a given meridian twice in succession.

Arc (37) – portion of a circle through which the celestial objects rise in the east and set in west, from the observer's point of view on Earth's surface.

Arid Climate (105) – dry (deficit moisture) area of land, having greater potential evapotranspiration than precipitation for a majority of months in a year.

Astronomical Unit (45) – mean distance of the Earth from the Sun.

Atmosphere (26, 77, 79) – thin shell of gases surrounding the Earth, separated (stratified) into layers each having distinct characteristics; troposphere, stratosphere, mesosphere, and thermosphere.

Atmosphere, Selected Properties of Earth's (215) – Reference Table.

Atmospheric Transparency (79, 85) – condition under which the Earth's atmosphere scatters, reflects, or absorbs the Sun's rays.

Atmospheric Variables (77) – weather changes in the atmosphere, including temperature, winds, moisture, air pressure, etc.

Banding (134) – pattern of layers caused by differences in the crystal alignments of various minerals in many metamorphic rocks; type of foliating.

Barometer (79) – instrument used to measure air pressure.

Bedrock (169, 190) – (local rock), solid rock underneath soil or exposed rock at Earth's surface.

Bedrock, Generalized Geology in NYS (203) – Reference Table.

Bench Mark (150) – marker in the ground indicating the exact elevation of that location above sea level.

Calorie (59) – heat quantity unit; amount of energy required to raise the temperature of 1 g of water through 1° C.; a **large calorie** (kilocalorie) is 1000 calories.

Capillarity (99) – upward movement of water against the force of gravity in a narrow space, such as a tube, plant vessel, or fine sand particles.

Capillary Water (99) – water that is found in the small spaces between fine grains of rock, sand, clay, or soil.

Carbon – 14 Dating (174) – process for determining the absolute age of a fossil or other material containing the element Carbon–14 (a radioactive isotope of Carbon–12 with a half–life of 5,600 years).

Carbonation (116)– the process of carbonic acid reaction with other materials.

Celestial Object (37) – "heavenly bodies," any object observed in the area above the Earth's atmosphere, including the Sun, Moon, stars, comets, planets, etc.

Cementation (136) – process in some sedimentary rocks, in which various sized sediments are cemented (glued) together by the action of precipitated minerals, resembles man–made concrete.

Centimeter Scale (201) – Reference Table.

Centrifugal Force (25) – outward force acting on a body rotating in a circle around a central point (focus).

Chemical Composition of Earth's Crust (211) – Reference Table.

Chemical Weathering (116) – process that alters the chemical characteristics of rocks and minerals, such as oxidation and hydration.

Circumference of Earth (26) – see Earth Dimensions.

Classification System (6) – organized data, based on observable properties.

Climate (105, 189) – discounting local weather changes, the average or normal weather of a particular large Earth area.

Cloud (88) – type of aerosol in the atmosphere composed of suspended small water droplets and/or ice crystals.

Cold Front (81) – the interface, leading edge, of an air mass which has cooler temperatures than the preceding warmer air mass, usually associated with moisture and precipitation.

Colloid (125) – small particles, from 10^{-4} to 10^{-6} millimeters across, which tend to remain in solution for long periods of time.

Compression (136) – process involved in the production of some sedimentary and metamorphic rocks, in which the weight of deposited sediments, water, and/or Earth movements presses underlying sediments together.

Compressional Wave (Primary, P–wave) (154) – seismic wave (action is like the expansion and contraction of a spring) which travels at a speed of thousands of kilometers per hour, through solids, liquids, or gases.

Condensation (57, 78, 87) – change in state from vapor (gas) to liquid, the loss of water vapor from warmer air onto a cooler surface, such as the condensation (fog) on a ice water glass.

Condensation Surface (or) Nucleus (87) – solid surface onto which water vapor may condense to form liquid droplets, in the formation of clouds – water condenses on dust or other particles, in the formation of early morning dew – water condenses on any solid and cooler surface.

Conduction (56, 85) – transfer of energy, usually heat, by contact from one atom to another atom within a liquid, gas or solid.

Conservation of Energy (59) – energy may be transformed, but may not be either created nor destroyed.

Constructional Forces (188) – see uplifting forces.

Contact Metamorphism (168) – the process of rock changing due to contact with hot magma or lava.

Continental Climate (106) – average weather of a land mass, little affected by large bodies of water, characterized by extremes in temperature.

Continental Crust (153, 156) – thick, low density upper part of the lithosphere that makes up the blocks (land mass) of a continent.

Continental Drift (153) – theory, backed by continual evidences, that continents (Earth plates) are now, as well as in the past, shifting positions.

Continental Polar Air Mass (80) – cP, usually a cold air mass originating in the land polar regions, such as in the northern most regions of Canada.

Continental Tropical Air Mass (80) – cT, warm air mass originating over warm land regions.

Contour Interval (31) – the difference in elevation between two consecutive contour lines.

Contour Line (31) – type of isoline on a topographic (contour) map, representing equal points of elevation.

Contour Map (31) – a topographic map, used as a model indicating elevations of the Earth's surface with the use of contour lines and symbols.

Convection (56) – transfer of energy, due to differences in substances' densities, in gases and liquids.

Convection Cell (current) (56, 86, 157) – circulatory motion in which heat energy is transferred from one place to another, due to density differences.

Convergence (86) – interfacing of air masses at the Earth's surface, in upper regions of the troposphere, making "air streams," vertical currents.

Coordinate System (27) – system or group of defined lines (may be imaginary lines) used for the determination or location of point(s) on a surface (such as graph, Longitude & latitude).

Coriolis Effect (41) – path of an object (or fluid) at the surface of the Earth undergoing a predictable horizontal deflection.

Correlation (169, 170) – match up of rock ages and geologic events.

Crust (149, 156) – layer of granite or basalt rock forming the outer part of the Earth's lithosphere.

Crystal (134, 135) – Earth material having a repeating pattern of characteristic shapes, due to a material's internal atomic structure, such as a cube (halite) and a tetrahedron (silicate).

Crystalline Structure (134, 139) – definite atomic pattern within a mineral (see crystal).

Crystallization (138) – formation of solid crystals into a rock, such as igneous rock, when the crystals separate from a magma solution.

Cycles (16, 46) – usually an orderly manner in which events in time and space repeat.

Cyclic Energy Transformation (46) – alternating between changes of energy from kinetic to potential and potential to kinetic; as seen in the Earth's changes of energy and orbital speed around the Sun.

Cyclone (80) – **LOW**, low pressure air mass with counterclockwise winds in the northern hemisphere, including violent weather, such as tornadoes and hurricanes.

Daily Motion (37) – apparent motion in an arc path across the Earth's sky from east to west during each 24 hour period.

Deficit (104) – see water budget; the local condition when the actual evapotranspiration is not equal to the potential evapotranspiration, due to insufficient precipitation and water soil storage; see drought.

Density (7, 8) – mass of a material divided by the material's volume.

Density Variables (86) – characteristics such as temperature and pressure that causes changes in density and phase changes.

Deposition (125) – settling out of solution of sediments and minerals in an erosional system.

Desert (105) – see water budget; arid area where the actual evapotranspiration is very much less than the potential evapotranspiration for that area.

Destructional Forces (188) – see leveling forces.

Dew Point Temperature (78 88, 89) – temperature at which water vapor present in the air saturates air and begins to condense; dew forms.

Dew Point Temperatures (212) – Reference Table.

Dimensional Quantities (7) – time, length, or mass.

Direct (vertical) Rays (66) – rays of solar energy hitting the surface of the Earth at an angle of 90°, also called perpendicular rays.

Displaced Sediments (118) – rock and mineral particles that are removed from their source and transported by water/wind to another place.

Distorted Structure (134) – resulting rock formations caused by Earth forces, such as heat and pressure, which bend, break, and fold rock layers.

Divergence (86) – following air mass convergence, ascending or descending, the spreading apart of air currents.

Drainage Patterns (191) – dendritic (random drainage over bedrock), trellised (parallel folds and faults), and radial (volcanic cones, young domes).

Drought (104) – a prolonged period of deficit weather conditions.

Duration of Insolation (66) – length of time that the Sun's rays are received at a particular location on the Earth's surface; varies with latitude and season; Earth surface temperature is directly proportional to the duration of insolation.

Dynamic Equilibrium (84, 128, 188) – a balance between two opposing processes going on at the same rate in a system, such as erosion and deposition and evaporation and condensation; refers to a landscape as well.

Earth Dimensions (23) – includes the Earth's circumference, radius, diameter, volume, and surface area.

Earth Energy (56) – secondary energy (solar is primary) such as the natural decay of radioactive matter on the Earth.

Earth's Interior, Inferred Properties of (210) – Reference Table.

Earthquake (151, 154) – sudden trembling or shaking of the ground, usually caused by a shifting of rock layers along a fault or fissure under the Earth's surface.

Earthquake P-wave and S-wave Travel Time (211) – Reference Table.

Electromagnetic Energy (53) – any energy radiated in transverse wave form, such as radio, sound, light, X–rays, etc.

Electromagnetic Spectrum (54) – wide range of wavelengths from lower frequencies such as radio waves short wave, AM , FM, TV, and radar to mid frequencies such as infrared rays, visible light, and ultraviolet, to high frequencies such as X–rays and gamma rays.

Electromagnetic Spectrum (214) – Reference Table.

Ellipse (44) – flattened circular path, having two foci (fixed radii); typical of the orbits of most all celestial objects and the Earth.

Elliptical Eccentricity (44) – degree of the "out of roundness" of the ellipse, as determined by the distance between the two foci divided by the length of the major axis of the ellipse.

Energy (46, 53, 59, 128) – ability to do work; forms: kinetic and potential.

Environmental Equilibrium (17, 188) – general stable and balanced state of the environment, changeable easily on a small scale.

Epicenter (155) – point on Earth's surface that is directly above the focus of an earthquake.

Equations and Proportions (216) – Reference Table.

Equilibrium, State of (17) – the tendency to remain unchanged.

Equinox (40, 66) – time at which the Sun's rays are directly perpendicular to the Earth's equator; equal day and night on the Earth; usually March 21st and September 23rd.

Erosion (118, 119, 172) – altering of the Earth's surface by the removing of rock, soil, and mineral pieces from one location to another by the action of water (liquid or solid) or wind.

Erosional–Depositional System (125, 127) – the system involving the opposing processes of erosion and deposition, involving energy relationships and dynamic equilibrium.

Error (7) – difference between the actual and observed measurements.

Escarpment (190) – steep slope separating two gently sloping surfaces.

Evaporite (137) – form of sedimentary rock, caused by the precipitation of minerals from evaporating water, such as, limestone, dolostone, gypsum and salt.

Evaporation (57, 883, 137) – change of phase from liquid to vapor (gas) occurring at the surface of that liquid.

Evapotranspiration (84) – combination of both processes evaporation and transpiration.

Event (15) – the occurrence of a change in the environment.

Extrusion (139) – mass of hardened lava at the Earth's surface, a type of igneous rock formation.

Extrusive Igneous Rock (139) – igneous rock that forms by the hardening of magma (hot liquid rock beneath the Earth's surface) after reaching the surface of the Earth.

Fault (150, 188, 190) – crack in the crust of the Earth along which rocks have moved.

Field (29) – region of space which contains a measurable quantity at every point.

Focus (geometric definition) **(44)** (pl.– foci) – fixed point from which a radius of 360° is a circle or sphere; two fixed points (foci) are required to produce an ellipse or oblate spheroid.

Focus (earthquake) **(154)** – point of origin of an earthquake.

Folded Strata (150, 190) – bend in the rock strata produced during the mountain– building process.

Fossil (150, 169, 170) – remains or traces of a once–living organism in sedimentary rock.

Foucault Pendulum (41) – freely swinging pendulum, which when allowed to swing without interference appears to change direction in a predictable manner due to the Earth's rotation.

Frames of Reference (15) – way to describe an environmental change using time and space.

Frequency of Waves (54) – cycle of differing waves; see electromagnetic spectrum.

Friction (58) – force found at the contact of two surfaces that offers resistance to motion, often producing heat or another form of energy.

Frictional Drag (85) – in the atmosphere, the slowing down of wind at the interface of the Earth surface caused by friction and the Coriolis effect.

Front (81) – the interface between two different air masses, such as the point of contact between warm and cold air masses.

Geocentric Model (38, 43) – an early attempt to explain the motions of celestial objects using the Earth as the stationary center for the orbiting celestial objects.

Geographic Poles (28) – actual axis points on the Earth, north and south, on which the Earth rotates.

Geologic History of NYS (208 – 209) – Reference Table.

Geologic Time Scale (170, 172) – geological periods, a scale of time that serves as a reference for correlating various events in the history of the Earth, divided into three main groups – eras, periods, and epochs, based on the study of rock history.

Geosyncline (151, 158) – shallow ocean basin along a continent where marine deposition occurs.

Graded Bedding (126) – layering of sediment in a fashion where heavier and/or largest particles are on the bottom and lighter and/or smaller particles are on top, in decreasing size.

Gradient (30) – expression of the degree of change of a field quantity from place to place; may also be referred to as an average slope.

Gravitational Force (24, 46, 119) – or gravity; attraction between any two objects in the universe.

Greenhouse Effect (58, 70) – process which increases the atmospheric temperature of the Earth, due to the transmission of short wave radiation through the atmosphere, absorption and conversion to long wave radiation at the Earth's surface.

Ground Water (98) – water that is found under the Earth's surface as the result of infiltration and storage.

Half–life (173) – time taken for half of a radioactive material to decay to its stable decay product; time for half of the atoms present to disintegrate.

Heat Energy (57, 59) – energy transferred due to the differences of temperature between two substances.

Heat of Fusion (57) – amount of latent heat involved in melting or freezing.

Heat of Vaporization (57) – amount of latent heat involved in evaporation or condensation.

Heliocentric Model (38, 44) – a modern attempt to explain the motions of celestial objects using the Sun as the stationary center for the orbiting celestial objects and fixed star positions.

HIGH (81) – see anticyclone, high pressure air mass, clockwise rotation in the northern hemisphere, more dense than a LOW.

High Noon (40) – 12:00 (noon) point in time when the Sun is at its highest altitude (zenith) on the observer's meridian.

Horizontal Sorting (126) – sorting of sediments in a stream by decreasing velocity, larger particles first laid down, followed by smaller and smaller particles down stream.

Humid Climate (105) – see water budget, an area where the precipitation is greater than the potential evaporation for the majority of months in a year.

Hydration (116) – the chemical reaction of water with other materials.

Hydrosphere (27) – thin layer of water which covers a majority (71%) of the Earth's surface.

Igneous Rock (138, 140) – rock formed by the solidification and crystallization of magma or lava (hot molten rock).

Igneous Rock Identification (206) – Reference Table.

Incident Insolation (69) – point and time of solar radiation hitting the Earth's surface; see angle of insolation and insolation.

Index Fossil (170) – fossil that is characteristic of a certain geologic time, sometimes referred to as a guide fossil.

Inferences (5) – interpretations (conclusions) based on observations.

Infiltration (99) – seeping and absorption of water into ground storage.

Inner Core (156) – iron/nickel solid inner sphere (zone) of the Earth's interior.

Insolation (66, 67, 78, 85) – **Incoming solar radiation**; Sun's energy that transmits through Earth's atmosphere and reaches Earth's surface.

Instruments (5) – tools used by the observer to improve on detail or extend the ability to obtain information and measurements.

Intensity of Insolation (66) – rate (amount and duration) of solar radiation reaching the Earth's surface.

Interface (17) – boundary between materials at which a change in environmental equilibrium occurs involving a loss or gain in energy states.

Intrusion (139) – rock mass formed from liquid rock (magma) cooling below the Earth's surface, igneous rock.

Intrusive Igneous Rock (139) – igneous rock formed below the Earth's surface by the hardening of magma (hot liquid rock).

Isobar (30, 79) – type of isoline on a weather map used to indicate equal air pressure points.

Isoline (30) – line representing equal values on a map or model (such as contours, isotherms, isobars) of field characteristics in two dimensions.

Isostasy (152, 158) – condition of equilibrium in the Earth's crust in which masses of greatest density are lower than those of lesser density.

Iso–surface (30) – model representing field characteristics in 3 dimensions.

Isotherm (30) – type of isoline on a weather map used to indicate equal temperature points.

Isotope (173) – a variety of an element that has the same atomic number, but a different atomic mass, due to a difference in the number of neutrons present in the nucleus, used for correlation studies when radioactive.

Jet Stream (83) – wavelike currents with high winds at upper levels which tend to control storm tracks.

Joint (168) – crack in a rock mass or rock where unlike a fault, no vertical or horizontal displacement has occurred.

Kepler's Harmonic Law (45) – explanation of planetary motion; relates a planet's period to its distance from the Sun.

Kinetic Energy (46) – energy of action, motion or at work.

Landscape (185, 186, 187) – topography of the land, inc. the characteristics of the Earth's surface.

Landscape Regions of NYS (202) – Reference Table.

Lapse Rate (213) – Reference Table.

Latent Heat (57) – energy released or absorbed during a phase change, such as a liquid to a gas, but with no temperature change involved.

Latitude (24, 28, 105) – distance north or south of the equator measured in degrees (parallels) from 0° at the equator, to 90° at the geographic poles.

Latitudinal Climate Pattern (105) – zones, West to East belts, of long term weather, primarily due to factors of temperature, winds, ocean currents, moisture, etc.

Lava (138) – magma that reaches the Earth's surface.

Length (7) – distance between the ends or sections of an object; usually measured in meters, centimeters, of millimeters.

Leveling Forces (Destructional) (188) – forces of weathering, erosion, transportation, deposition, and subsidence.

Lithosphere (27, 149) – continuous outer solid rock shell of the Earth.

Local Water Budget (103, 105) – system of accounting for an area's water yearly supply, see water budget.

Long Waves or L–Waves (154) – waves that travel along the Earth's surface.

Longitude (28, 29) – distance east or west of the prime meridian measured in degrees from 0° at the prime meridian (runs through Greenwich, England – Greenwich Mean Time) to 180° east or west (in the Pacific Ocean – International Date Line).

LOW (80) – see cyclone, a low pressure air mass, counterclockwise rotation in the northern hemisphere, less dense than a HIGH.

Magma (138) – molten rock material beneath the Earth's surface.

Magnetic Polarity (152) – see Mid–Oceanic Ridge.

Major Axis (44) – the longest diameter of an ellipse

Mantle (156) – layer of the Earth between the crust and the core.

Mantle Convection Cells (157) – the movement of heat and matter caused by differences in density within the Earth's mantle.

Map Scale (31) – the ratio between the distance on a map and the distance on the Earth's surface.

Marine Climate (105) – long term weather characteristics of an area near water bodies, such as large oceans characterized by small seasonal temperature ranges and abundant precipitation.

Maritime Polar Air Mass (80) – mP – a cool, moist air mass originating over a cold water surface.

Maritime Tropical Air Mass (80) – mT – usually a warm air mass originating over tropical waters, such as the Caribbean region.

Mass (6) – amount (quantity) of matter which an object contains.

Mathematical Combinations (7) – in measurement, combinations of basic dimensional quantities, density, pressure, volume, acceleration, etc.

Mean Solar Day (42) – average solar day based on the apparent solar day which is constantly changing due to the speed changes and rotation of the Earth.

Meander (127) – curving pattern of a river due to erosion and deposition.

Measurement (6, 71) – use of time, length, or mass as a basic dimensional quantity (numerical).

Meridians (28) – grid "lines" running between the North Pole and the South Pole; used to measure longitude.

Metamorphic Rock (139, 140, 168) – rocks formed by the effect of heat pressure and/or chemical action on other rocks, a recrystallization of pre–existing rocks.

Metamorphic Rock Identification (207) – Reference Table.

Mid–Ocean Ridge (152) – mountain ridge in mid–ocean, such as the Mid–Atlantic Ridge, which extends for about 64,000 kilometers roughly parallel to continental margins.

Mineral (135) – inorganic (nonliving) crystalline, solid substance with a definite chemical (atomic) shape and composition.

Model (29) – description or representation of an idea or concept which helps to illustrate actions or information (for example, models can be used to illustrate the Earth's shape and size).

Moisture (78) water in any form in the atmosphere and other places.

Moisture Capacity (78, 104) – amount of water that can be held by an air mass, cold air generally hold less water than warm air, see absolute humidity; amount of moisture that can be held by soil.

Monomineralic Rock (135) – rock composed of just one mineral type.

Moon (39) – Earth's satellite.

Mountain (106, 186) – elevated landscape with distorted rock structure.

Nonsedimentary Rock (138) – rock type not formed from a sedimentary process, includes igneous and metamorphic rocks

Noon (40) – point in time when the Sun is directly on observer's meridian.

Nuclear Waste (102) – long–term harmful by–products of nuclear reactions.

Oblate Spheroid (23) – slightly flattened sphere; shape of the Earth, flattened at the geographic poles and bulging at the equator.

Observation (1, 185) – use of senses in measuring and collecting data concerning environment.

Occluded Front (83) – interface formed when a cold front overtakes a warm front.

Ocean–Floor Spreading (152) – theory supported by past and present evidence that the ocean floor is moving outwards from the mid–ocean ridge.

Ocean Currents, Surface (204) – Reference Table.

Oceanic Crust (152, 156) – thinner, more dense part of the Earth's crust composed of basaltic material.

Orbit (38, 44) – path of a celestial object, satellite, and/or Earth about a center, usually an ellipse.

Orbital Velocity (Speed) (45) – the speed of an object at any given time in its orbit; usually changing due to distances from its gravitational center.

Organic Evolution (174) – theory of change, an explanation of how new species develop by punctuated (rapid) or gradual (slow) changes.

Organic Substance (134) – material containing the element Carbon, usually associated with living or once living things.

Orographic Effect (106) – effect that mountains have on weather and climate; blockage of precipitation from the leeward side of mountains.

Outcrop (169) – exposed bedrock.

Outer Core (156) – liquid Earth zone between the inner core of the Earth and the mantle, like the inner core composed of iron and nickel.

Oxidation (116) – the chemical reaction of oxygen with other materials.

Parallels (28) – grid lines on a map or globe; another term used to determine latitude.

Parallelism of the Axis (67) – the Earth's axis at any place in its orbit is parallel to the axis in any other place in the Earth's orbit.

Particle Size (Transported) to Water Velocity Relationship (206) – Reference Table.

Percent Error (7) – or deviation, mathematical expression of a calculated error in percent (%).

Perihelion (40, 45) – point on the Earth's orbit when it is closest to the Sun; a distance of 147 million kilometers, usually occurring January 1st.

Period of Revolution (45) – amount of time an object takes to make one complete orbit around its center, in the case of the Earth, 365+ days to orbit the Sun.

Perpendicular Insolation (71) – vertical rays of the Sun; 90° radiation at Earth's surface.

Permeability (99) – rate at which moisture passes through a material.

Permeability Rate (99) – speed at which water pass through a porous material, see permeability and porosity.

Phase Change (8, 57) – change of a material through states of solid, liquid and gas.

Phases of the Moon (39) – changes in the amount of the illuminated surface of the Moon as seen from the Earth, cyclic over a 29 1/2 day period.

Physical Constants (216) – Reference Table.

Physical Weathering (115) – the process that alters the physical characteristics of rocks/minerals, generally leading to breaking into smaller pieces.

Plain (186) – low elevation landscape, gentle slopes and relatively stable, often composed of horizontal layers of sedimentary rocks.

Planetary Motions (38) – the non uniform movement of planets.

Planetary Period (45) – the time it takes a planet to make one revolution arount the Sun.

Planetary Wind Belts (87) – zones on the Earth where winds generally blow in one direction only, such as the prevailing southwest winds of the U.S.

Radioactive Dating (167) – process of determining the age of rock by measuring the half–life of radioactive materials in the rock.

Radioactivity (Decay) (56, 173) – secondary source of energy for the Earth; spontaneous and natural nuclear breakdown from unstable to stable atomic forms, energy is released, the process has a constant and predictable rate, not affected by environmental changes.

Random Reflection (69) – reflection (see scattering) of insolation due to aerosols in the atmosphere; dust and water droplets increase the amount of random reflection, causing a decrease in the amount of insolation reaching the Earth's surface.

Radiative Balance (71) – average energy levels on the Earth remain fairly constant due to the Earth giving off as much energy as it receives.

Recharge (104) – see water budget; the replacement of water by infiltration into the soil storage area.

Recrystallization (139) – formation of new crystalline materials by the enlargement of preexisting crystals by the action of thermal metamorphism.

Reflection (54, 69) – change in electromagnetic wave direction due to the non penetration of a wave into a surface; a smooth surface will reflect a wave at the same angle at which it strikes the surface.

Refraction (54) – change in direction of a material when an electromagnetic wave goes from one material to another solution with a different density.

Relative Age (Date) (167) – dates in the Earth's history determined with reference to other events helpful in determining relationships in a time line.

Relative Humidity (85) – ratio of the mass of water vapor per unit volume of the air to the mass of water vapor per unit volume of saturated air at the same temperature.

Relative Humidity (212) – Reference Table.

Residual Sediment (118) – sediment that remains at the site of weathering.

Reversal of Magnetic Polarity (152) – reference to changing polarity as observed in rocks because of the Earth's magnetic poles reversing.

Revolution (38, 41) – orbiting of one body around another body.

Rock Cycle (141) – model to explain the changes in rocks and the formation of sedimentary and nonsedimentary rocks, igneous and metamorphic rocks.

Rock Cycle in Earth's Crust (206) – Reference Table.

Rock Formation (141) – body or mass of rock with similar features and characteristics.

Rock Forming Minerals (135) – minerals, mostly silicates, that form 90% of the Earth's crust.

Rock Properties (133) – characteristics of various rocks.

Rock Resistance (190) – characteristic of rock types to resist the forces of change, including weathering and erosion.

Rotation (38, 41) – spinning of an object about its own axis.

Runoff (100) – water that does not infiltrate the soil storage area and flows over the land surface to lakes, streams, and oceans.

Satellite (39) – any object which is held by another object's force of gravitation, around which it revolves (orbits); the Moon is the Earth's satellite; a man–made object which orbits the Earth.

Saturation Point (84) – point at which the air is completely filled with water vapor to the air's maximum capacity, after which water will condense.

Saturation Vapor Pressure (84) – amount of vapor pressure of an air mass when filled to capacity with water vapor, 100% relative humidity.

Scalar Quantity (30) – a measurement, such as temperature or barometric pressure; a single field quantity having magnitude only.

Scattering (55, 69) – wave movement in different directions due to reflection.

Seasons (66) – divisions of the year caused by climatic changes, angle of insolation, Earth tilt, etc.; generally, spring, summer, fall, and winter.

Sediment (115, 133) – rock particles that are produced and/or transported by erosion and weathering.

Sediment Laden Flow (126) – movement of an erosional transport agent containing some forms of sediment, such as a glacier, turbidity current.

Sedimentary Rock (136) – rock formed from compaction and cementation of sediment.

Sedimentary Rock Identification (207) – Reference Table.

Sedimentation (125) – settling out of solution of sediments, inc. minerals, in an erosional system; deposition.

Seismic Wave (154) – wave that radiates from the point of origin of an earthquake, moving in all directions through solid rock.

Seismograph (154) – very delicate instrument that detects and records passing earthquake waves.

Senses (1) – sight, hearing, touch, smell, taste.

Settling Rate (125) – time required for a certain sediment to settle out of water or air.

Shear Wave (Secondary, S-wave) (154) – wave that causes individual rock particles to vibrate at right angles to the direction that the wave is traveling; cannot pass through liquids.

Silicon–Oxygen Tetrahedron (135) – structural model of a silicate pyramid.

Simple Celestial Model (44) – modern model of the Earth and celestial objects system, in which there is a center to the universe about which all objects revolve and a solar system much the same as in the heliocentric model.

Sink (59) – portion of an energy system with lower energy concentrations, into which energy usually flows.

Soil Association (185) – unit of soil classification, including the characteristics of the soil, composition, porosity, permeability, structure, and the ability to support life.

Soil Formation (117) – production of soil, particles of rocks and minerals, and organic matter.

Soil Horizons (117) – layers of soil produced as a result of the weathering processes and biologic activity; a) horizontal topsoil, b) horizontal subsoil.

Soil Storage (104) – amount of water held below Earth's surface in the soil.

Soil Storage Change (104) – see water budget; the amount of water either removed or added to a soil storage area; usage and recharge, respectively.

Solar Energy (55, 65) – any energy forms radiated from the Sun.

Solar Electromagnetic Spectrum (55) – full range of wavelengths emitted from the Sun with the maximum intensity occurring in the visible region,

Solar System (38, 44) – orbiting system of the Earth, planets, and moons with the Sun as the center of revolution.

Solar System Data (214) – Reference Table.

Solidification (138) – see crystallization.

Solstice (40) – times when the Sun's rays are perpendicular (at zenith) to the $23\frac{1}{2}°$ north latitude (about June 21st), summer solstice and $23\frac{1}{2}°$ south latitude (about December 21st), winter solstice.

Sorted and Unsorted Particles (126) – selection of various materials, based on size, the more similar the particle sizes, the greater the sorting, the greater the difference in the particle sizes, the less the sorting.

Sorting of Sediments (126) – manner in which materials in suspension settle out of a transport medium in a definite pattern.

Source (59) – portion of an energy system with the highest energy concentrations, from which energy usually flows.

Source Region (80) – place on the Earth where an air mass forms.

Species (174) – most specific part of the classification system, or two organisms of the same species are able to mate and produce fertile offspring.

Specific Heat (59) – amount of heat necessary to raise the temperature of 1 g of any substance 1° C, measured in calories; specific heat of water is 1; most other substances have specific heats of less than 1.

Station Model (89) – on a weather map, describes weather conditions at a reporting station.

Station Model, Weather Maps (213) – Reference Table.

Stationary Front (83) – interface of two air masses that do not move.

Strata (150) – layers of rock material, usually sedimentary rock.

Stream Bed (120) – interface of the water and bottom of a stream, including the rock particles and bottom materials.

Stream Discharge (104) – measurement of amount of water passing a certain point in a stream in a certain amount of time; rate of flow in volume.

Stream Drainage Pattern (191) – pattern that forms due to the way water drains across the land in a stream or river system.

Sublimation (87) – phase change from a solid to a gas or a gas to a solid without a liquid phase, as in the sublimation of ice to water vapor without melting to a liquid form.

Subsidence (150) – act of sinking or settling of the Earth's surface.

Sun (39) – star, center of the solar system.

Surface Water (100) – water on the surface of the Earth.

Surplus (104) – see water budget, surface water that neither evaporates nor infiltrates, but is runoff water.

Synoptic Analysis (80) – short discussion of major factors; summary.

Technology (191) – science of the means and scientific discoveries employed to provide objects necessary for human sustenance and comfort.

Tectonics (157) – study of the Earth's crustal movements usually concerned with the folding and faulting associated with mountain building.

Tectonic Plates (205) – Reference Table.

Temperature (57, 68) – measurement of the average kinetic energy of a substance.

Temperature Scales (213) – Reference Table.

Terrestrial Motions (41) – revolution and rotation of the Earth.

Texture (139) – characteristics of rocks including size, shape, and particle arrangement.

Tilted Strata (150) – layers of sedimentary rock that have been moved from a horizontal to an angular position by the action of Earth movements.

Time (7, 15) – unit of measurement that depends upon the rotation of the Earth and its relative position to the perpendicular Sun's rays; on the Earth the standard of time is the Greenwich Mean Time; the International Date Line is 12 hours before or after.

Topographic Map (31) – see contour map.

Track (83, 107) – in weather conditions, the often predictable path that an air mass or a front takes.

Transmission (55) – process of energy moving through atmosphere unchanged.

Transition Zone (141) – interface area within a rock mass where one form of rock changes to another form, such as sedimentary to metamorphic.

Transpiration (84, 104) – process by which plants release water vapor through leaf and stem openings into the atmosphere.

Transported Sediment (118) – materials produced by weathering and organic materials which are moved from their place of origin by the agents of erosion.

Transporting Agents (118) – forces that affect erosion and move sediments from one place to another, such as water and wind.

Transporting System (118) – all agents involved in erosion and movement, including erosion, the transporting agent, and the material moved.

Transverse Wave (53) – right angle vibrations of electromagnetic waves from the source of the waves.

Unconformity (168) – zone where rocks of different ages meet, representing missing rocks, or a gap in the rock history; a buried erosional surface.

Uniformitarianism (172) – geologic principle that states that processes of the present are similar to processes of the past geologic ages.

Uplifting (Constructional) Forces (188) – Earth changing forces, including plate tectonics, isostasy, volcanoes, continental lift, earthquakes, and ocean–floor spreading.

Uranium – 238 (173) – radioactive isotope of uranium with a half–life of 4.5 billion years, used in the dating of very old materials.

Usage (104) – see water budget, the loss of water through evapotranspiration from the soil storage area.

Vapor Pressure (84) – pressure exerted by water vapor (gas) within a specific volume of air, directly related to absolute humidity.

Variable (16) – a factor involved in a change; changeable part of experiment.

Vector Field (30) – field representing both a magnitude and a direction, such as gravitational and magnetic fields.

Vector Quantity (30) – quantity (amount) having both magnitude and direction, such as wind velocity.

Vein (168) – narrow well–defined zone containing mineral–bearing rock in place.

Visible Light Spectrum (55) – portion of the electromagnetic spectrum between infrared and ultraviolet rays, including the colors red, orange, yellow, green, blue, indigo and violet.

Volcanic Ash (170) – fine particles of igneous rock ejected from an active volcano during a volcanic eruption, dating the age of the ash gives the relative age of the volcanic eruption, since the ash was produced during the eruption.

Volume (7) – combination of three dimensions – length, width, and height.

Walking the Outcrop (169) – the process of physically tracing a rock layer from one location to another.

Warm Front (81) – interface, leading edge, of an air mass which has warmer temperatures than the preceding air mass, usually associated with steady precipitation.

Water Budget (103) – see local water budget, the expression of the amount of water an area gains in precipitation and loses by evapotranspiration.

Water Cycle (197) – phase changes and the movements of water between atmosphere, hydrosphere, and lithosphere; hydrologic cycles.

Water Purification (97) – the cleaning of water through the water cycle, precipitation and evapotranspiration.

Water Table (99) – top of the Zone of Saturation in the ground.

Water Vapor (84) – state of water in which it is a gas.

Water Velocity Relationship to Transported Particle Size (206) – Reference Table.

Wavelength (54) – distance between the peaks (crests) of two successive waves; the closer the waves (shorter the wavelength) the stronger and higher frequency the waves.

Weather (76, 80) – short term (hours or days) conditions of the atmosphere determined by variables such as temperature, wind, moisture, and pressure.

Weather Map Information and Symbols (213) – Reference Table.

Weather Maps (80, 89) – "bird's eye view" of Earth's weather patterns.

Weather Prediction (80) – forecasting of temperature, air pressure, moisture, and air movement changes.

Weathering (115, 117) – physical and chemical processes that break the rocks on the Earth's surface.

Weight (6) – a measurement of the pull of gravity on an object.

Wind (79, 86) – horizontal movement of air, often in currents, over the surface of the Earth.

Year (42) – amount of time required for the Earth to make one complete revolution around the Sun, about 365¼ days (24 hour rotations).

Zenith Point (40) – highest point of the arc of a celestial object, perpendicular to the Earth's surface over the observer's longitude (meridian).

Zone of Aeration (97) – ground above the water table.

Zones of Convergence, **Divergence (86)** – regions of LOW pressure and HIGH pressure, respectively.

Zone of Saturation (97, 99) – ground below the water table where the pores are filled with water.

Assignments:

Earth Science Practice Examination 1 — June 1993

Part I Answer all 55 questions in this part. (55)

Directions (1-55): For *each* statement or question, select the word or expression that, of those given, best completes the statement or answers the question.

page 241

1   The map at the right shows the elevation field for a 30-by-50-meter section of a parking lot on which a large pile of sand has been dumped. The isolines show the height of the sand above the surface of the parking lot in meters.

Which map represents the most likely elevation field for the same area after several heavy rainstorms?

(1)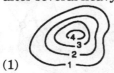

(3)

(2)

(4)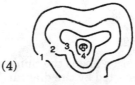

2   The Earth's actual shape is most correctly described as
    1   a circle   3   an oblate sphere
    2   a perfect sphere   4   an eccentric ellipse

3   According to the *Earth Science Reference Tables*, where are atmospheric pressure readings of 10^{-2} atmosphere found?
    1   stratosphere   3   mesosphere
    2   troposphere   4   thermosphere

4   The diagram at the right shows the altitude of Polaris above the horizon at a certain location.
    What is the latitude of the observer?
    (1) 10° N
    (2) 40° N
    (3) 50° N
    (4) 90° N

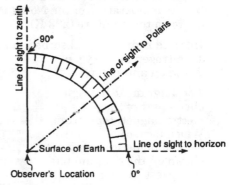

5   In the Northern Hemisphere, during which season does the Earth reach its greatest distance from the Sun?
    1   winter   3   summer
    2   spring   4   fall

6   The Coriolis effect provides evidence that the Earth
    1   has a magnetic field   3   revolves around the Sun
    2   has an elliptical orbit   4   rotates on its axis

7   During the warmest part of a June day, breezes blow from the ocean
    toward the shore at a Long Island beach. Which statement best explains
    why this happens?
    1   Winds usually blow from hot to cold areas.
    2   Winds never blow from the shore toward the ocean.
    3   Air pressure over the ocean is higher than air pressure over the land.
    4   Air pressure over the land is higher than air pressure over the ocean.

Base your answers to questions 8 and 9
on the diagram at the right which
represents the apparent daily path of the
Sun across the sky in the Northern
Hemisphere on the dates indicated.

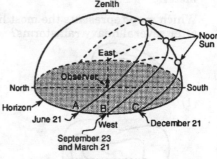

8   At noon on which date would the
    observer cast the longest shadow?
    1   June 21
    2   September 23
    3   March 21
    4   December 21

9   Which observation about the Sun's apparent path at this location on June
    21 is best supported by the diagram?
    1   The Sun appears to move across the sky at a rate of 1° per hour.
    2   The Sun's total daytime path is shortest on this date.
    3   Sunrise occurs north of east.
    4   Sunset occurs south of west.

10 At which temperature would an object radiate the *least* amount of
    electromagnetic energy?
    1   the boiling point of water (100°C)
    2   the temperature at the stratopause (0°C)
    3   the temperature of the North Pole on December 21 (-60°F)
    4   room temperature (293 K)

11 Infrared radiation is absorbed in the atmosphere mainly by
    1   nitrogen and oxygen   3   ice crystals and dust
    2   argon and radon   4   carbon dioxide and water vapor

12 The diagram at the right
    shows part of the
    electromagnetic spectrum.
    Which form of
    electromagnetic energy
    shown on the diagram has
    the lowest frequency and
    longest wavelength?
    1   AM radio
    2   infrared rays
    3   red light
    4   gamma rays

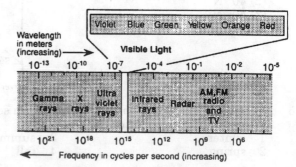

13 The factor that contributes most to the seasonal temperature changes during 1 year in New York State is the changing
1 speed at which the Earth travels in its orbit around the Sun
2 angle at which the Sun's rays strike the Earth's surface
3 distance between the Earth and the Sun
4 energy given off by the Sun

14 The tilt of the Earth on its axis is a cause of the Earth's
(1) uniform daylight hours
(2) changing length of day and night
(3) 24-hour day
(4) 365¼ -day year

15 Which process most directly results in cloud formation?
1 condensation   3 precipitation
2 transpiration   4 radiation

16 An air mass originating over north central Canada would most likely be
1 warm and dry   3 cold and dry
2 warm and moist   4 cold and moist

17 Which graph best represents the relationship between the moisture-holding capacity (ability to hold moisture) of the atmosphere and atmospheric temperature?

(1)

(2)

(3)

(4)

18 In the diagram of a mountain at the right , location *A* and location *B* have the same elevation.

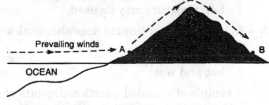

Compared to the climate at location A, the climate at location B will be
1 warmer and drier   3 warmer and wetter
2 cooler and drier   4 cooler and wetter

19 The graph at the right shows the air temperature and dewpoint temperature at one location at four different times during one morning.

At what time was the chance of precipitation the greatest?
(1) 1 a.m.
(2) 5 a.m.
(3) 3 a.m.
(4) 7 a.m.

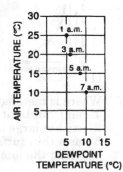

20   According to the *Earth Science Reference Tables*, what is the approximate
      dewpoint temperature when the dry-bulb temperature is 18°C and the
      wet-bulb temperature is 14°C?
      (1) 8.0°C   (3) 11°C
      (2) 10.°C   (4) 12°C

21   Which earth material covering the surface of a landfill would permit the
      least amount of rainwater to infiltrate the surface?
      1   silt   3   sand
      2   clay   4   pebbles

22   Two coastal cities have the same latitude and elevation, but are located
      near different oceans. Which statement best explains why the two cities
      have different climates?
      1   They are at different longitudes.
      2   They are near different ocean currents.
      3   They have different angles of insolation.
      4   They have different numbers of daylight hours.

23   Which graph best represents the relationship between soil permeability
      rate and infiltration when all other conditions are the same?

      (1)   (2)   (3)   (4)

24   What most likely will happen to soil moisture when precipitation is greater
      than potential evapotranspiration?
      1   Soil-moisture storage may decrease.
      2   Soil-moisture deficit may increase.
      3   Soil moisture may be recharged.
      4   Soil moisture may be used.

25   In which type of climate does chemical weathering usually occur most
      rapidly?
      1   hot and dry   3   cold and dry
      2   hot and wet   4   cold and wet

26   A sample of rounded quartz sediments of different particle sizes is dropped
      into a container of water. Which graph best shows the settling time for
      these particles?

      (1)   (2)   (3)   (4)

27   Where is metamorphic rock frequently found?
      1   on mountaintops that have horizontal layers containing marine fossils
      2   within large lava flows
      3   as a thin surface layer covering huge areas of the continents
      4   along the interface between igneous intrusions & sedimentary bedrock

28 The map at the right shows the top view of a meandering stream as it enters a lake. At which position along the stream are erosion and deposition dominant?

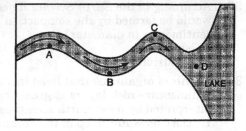

 1 Erosion is dominant at *A* and *D*, and deposition is dominant at *B* and *C*.
 2 Erosion is dominant at *B* and *C*, and deposition is dominant at *A* and *D*.
 3 Erosion is dominant at *A* and *C*, and deposition is dominant at *B* and *D*.
 4 Erosion is dominant at *B* and *D*, and deposition is dominant at *A* and *C*.

29 The particles in a sand dune deposit are small and very well-sorted and have surface pits that give them a frosted appearance. This deposit most likely was transported by
 1 ocean currents   3 gravity
 2 glacial ice   4 wind

30 The diagram at the right represents a sedimentary rock outcrop. Which rock layer is the most resistant to weathering?
 (1) 1
 (2) 2
 (3) 3
 (4) 4

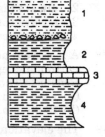

31 Which property is most useful in mineral identification?
 1 hardness   3 size
 2 color   4 texture

32 The diagrams below represent four rock samples. Which rock took the longest time to solidify from magma deep within the Earth?

Bands of alternating light and dark minerals
(1)

Easily split layers of 0.0001-cm-diameter particles cemented together
(2)

Glassy black rock that breaks with a shell-shape fracture
(3)

Interlocking 0.5-cm-diameter crystals of various colors
(4)

33 According to the *Earth Science Reference Tables*, which two elements make up the greatest volume of the Earth's crust?
 1 silicon and potassium   3 iron and nickel
 2 silicon and iron   4 oxygen and potassium

34 The recrystallization of unmelted material under high temperature and pressure results in
 1 metamorphic rock   3 igneous rock
 2 sedimentary rock   4 volcanic rock

35   According to the *Earth Science Reference Tables*, which sedimentary rock would be formed by the compaction and cementation of particles 1.5 centimeters in diameter?

     1   shale   3   conglomerate
     2   sandstone   4   siltstone

36   Fossils of organisms that lived in shallow water can be found in horizontal sedimentary rock layers at great ocean depths. This fact is generally interpreted by most Earth scientists as evidence that

     1   the cold water deep in the ocean kills shallow-water organisms
     2   sunlight once penetrated to the deepest parts of the ocean
     3   organisms that live in deep water evolved from species that once lived in shallow water
     4   sections of the Earth's crust have changed their elevations relative to sea level

37   Earthquakes generate compressional waves (*P*-waves) and shear waves (*S*-waves). Compared to the speed of shear waves in a given earth material, the speed of compressional waves is

     1   always slower
     2   always faster
     3   always the same
     4   sometimes faster and sometimes slower

38   A P-wave reaches a seismograph station 2,600 kilometers from an earthquake epicenter at 12:10 p.m. According to the *Earth Science Reference Tables*, at what time did the earthquake occur?

     (1) 12:01 p.m.   (3) 12:15 p.m.
     (2) 12:06 p.m.   (4) 12:19 p.m.

39   The elevation of a certain area was measured for many years, and the results are recorded in the data table at the right. If the elevation continued to increase at the same rate, what was most likely the elevation of this area in 1990?

     (1) 103.25 m   (3) 103.75 m
     (2) 103.50 m   (4) 104.00 m

Year	Elevation (m)
1870	102.00
1890	102.25
1910	102.50
1930	102.75
1950	103.00

40   The cross-sectional diagram below of the Earth shows the paths of seismic waves from an earthquake. Letter *X* represents the location of a seismic station. Which statement best explains why station *X* received only *P*-waves?

     (1) *S*-waves traveled too slowly for seismographs to detect them.
     (2) Station *X* is too far from the focus for *S*-waves to reach.
     (3) A liquid zone within the Earth stops *S*-waves.
     (4) *P*-waves and *S*-waves are refracted by the Earth's core.

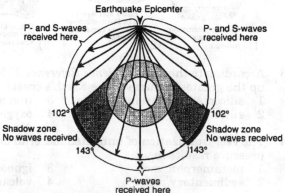

41 Geologists have subdivided geologic time into periods that are based on
1 carbon dating
2 rock types
3 fossil evidence
4 landscape regions

42 An ancient bone was analyzed and found to contain carbon-14 that had decayed for nearly two half-lives. According to the *Earth Science Reference Tables*, approximately how old is the bone?
(1) 1,400 years
(2) 2,850 years
(3) 5,700 years
(4) 11,400 years

43 According to the *Earth Science Reference Tables*, studies of the rock record suggest that
1 the period during which humans have existed is very brief compared to geologic time
2 evidence of the existence of humans is present over much of the geologic past
3 humans first appeared at the time of the intrusion of the Palisades sill
4 the earliest humans lived at the same time as the dinosaurs

44 The diagram at the right shows a cross-sectional view of part of the Earth's crust. What does the unconformity (buried erosional surface) at line *XY* represent?
1 an area of contact metamorphism
2 a time gap in the rock record of the area
3 proof that no deposition occurred between the Cambrian and Carboniferous periods
4 overturning of the Cambrian and Carboniferous rock layers

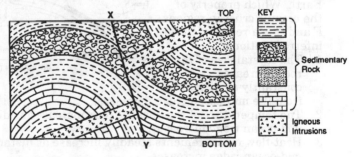

45 The diagram at the right shows a cross section of the Earth's crust. Line *XY* is a fault. Which sequence of events, from oldest to youngest, has occurred in this outcrop?

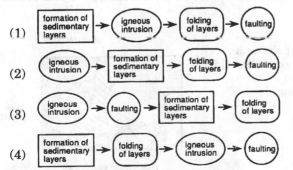

(1) formation of sedimentary layers → igneous intrusion → folding of layers → faulting

(2) igneous intrusion → formation of sedimentary layers → folding of layers → faulting

(3) igneous intrusion → faulting → formation of sedimentary layers → folding of layers

(4) formation of sedimentary layers → folding of layers → igneous intrusion → faulting

46 Which geologic evidence best supports the inference that a continental ice
    sheet once covered most of New York State?
    1   polished and smooth pebbles; meandering rivers; V-shaped valleys
    2   scratched and polished bedrock; unsorted gravel deposits; transported
        boulders
    3   sand and silt beaches; giant swamps; marine fossils found on
        mountaintops
    4   basaltic bedrock; folded, faulted, and tilted rock structures; lava flows

47 The boundaries between landscape regions are usually indicated by sharp
    changes in
    1   bedrock structure and elevation
    2   weathering rate and method of deposition
    3   soil associations and geologic age
    4   stream discharge rate and direction of flow

48 According to the *Earth Science Reference Tables*, which type of landscape
    region is found at 44° North latitude and 75° West longitude?
    1   plains   3   lowlands
    2   plateaus   4   mountains

49 The diagram at the right
    represents a partial cross
    section of a model of the
    Earth. The arrows show
    inferred motions within the
    Earth. Which property of
    the oceanic crust in regions
    F and *G* is a result of these
    inferred motions?
    1   The crystal size of the
        rock decreases
        constantly as distance
        from the mid-ocean ridge increases.
    2   The temperature of the basaltic rock increases as distance from the
        mid-ocean ridge increases.
    3   Heat-flow measurements steadily increase as distance from the
        mid-ocean ridge increases.
    4   The age of the igneous rock increases as distance from the mid-ocean
        ridge increases.

50 The diagram at the right shows a
    sample of conglomerate rock. The oldest
    part of this sample is the
    1   conglomerate rock sample
    2   calcite cement
    3   limestone particles
    4   mineral vein

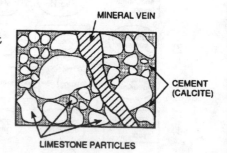

51 The map at the right shows the
    drainage patterns of a volcanic
    region. Which two locations are
    most likely volcanic mountain
    peaks?
    (1) *A* and *B*
    (2) *B* and *C*
    (3) *A* and *D*
    (4) *B* and *D*

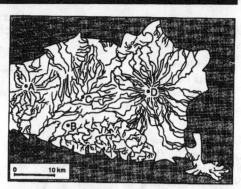

52 The map at the right shows
    average annual temperatures in
    degrees Fahrenheit across the United
    States. Which climatic factor is most
    important in determining the pattern
    shown in the eastern half of the
    United States?
    1   ocean currents
    2   mountain barriers
    3   elevation above sea level
    4   latitude

Note that questions 53 through 55 have only three choices.

53 The diagram below shows an area where sea level gradually dropped over a
    period of thousands of years. A continuous sandy beach deposit stretching
    from *A* to *B* was created.

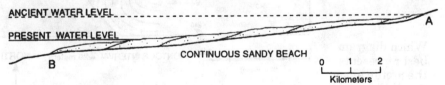

    Which statement about the beach deposit would most likely be true?
    1   It is older at *A* than at *B*.
    2   It is older at *B* than at *A*.
    3   It is the same age at *A* and *B*.

54 As warm, moist air moves into a region, barometric pressure readings in
    the region will generally
    1   decrease
    2   increase
    3   remain the same

55 As surface runoff in a region increases, stream discharge in that region will
    usually
    1   decrease
    2   increase
    3   remain the same

Part II

This part consists of ten groups, each containing five questions. Choose seven of these ten groups. [35]

Group 1
If you choose this group, be sure to answer questions 56-60.

Base your answers to questions 56 through 60 on the *Earth Science Reference Tables*, the contour map at the right, and your knowledge of Earth Science. Points *A*, *B*, *Y*, and *Z* are reference points on the map. Note that portions of the map are incomplete.

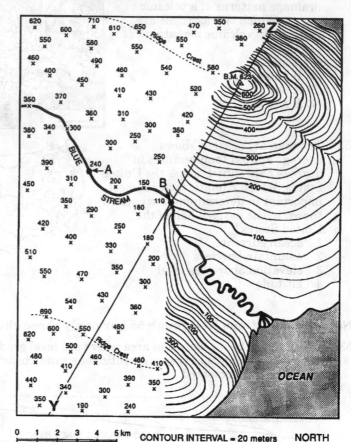

56   Which diagram best represents the profile along line *YZ*?

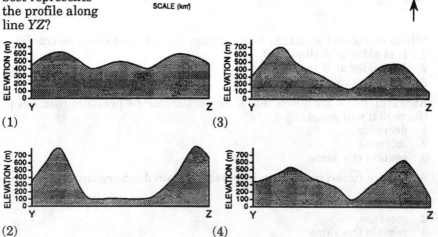

0 1 2 3 4 5 km   CONTOUR INTERVAL = 20 meters   NORTH

SCALE (km)

57 A benchmark (B.M. 623) located at the top of the hill is shown as ▲.
What is the elevation of the contour line that is closest to this benchmark?
(1) 600 m   (2) 610 m   (3) 620 m   (4) 630 m

58 In which general direction is Blue Stream flowing?
1   east   2   west   3   northwest   4   southeast

59 The gradient of Blue Stream between point *A* and point *B* is approximately
(1) 26 m/km   (3) 130 m/km
(2) 70. m/km   (4) 350 m/km

60 On which map is the 300-meter contour line completed correctly?

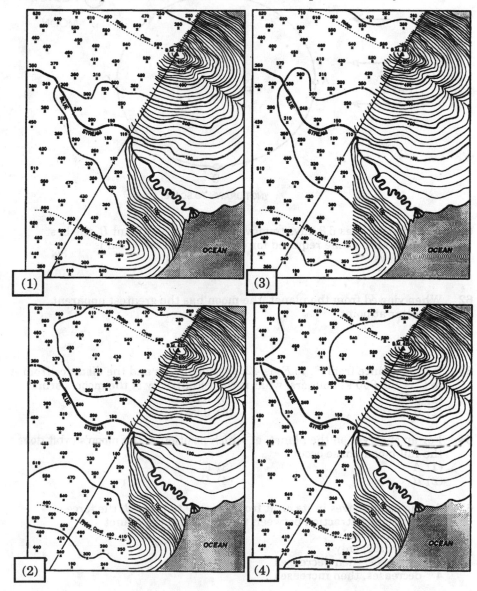

Group 2
If you choose this group, be sure to answer questions 61-65.

Base your answers to questions 61 through 65 on the *Earth Science Reference Tables*, the diagram below, and your knowledge of earth science. The diagram represents a model of the orbit of a moon around a planet. points A, B, C, and D indicate four positions of the moon in its orbit. Points F_1 and F_2 are focal points of the orbit.

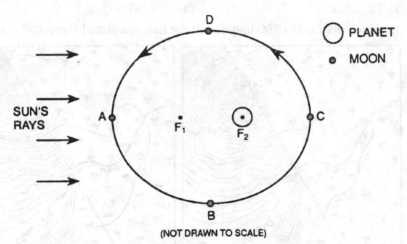

(NOT DRAWN TO SCALE)

61   If the moon takes 6.8 days to move from point A to point B, the best estimate of the time required for one complete revolution is
(1) 20. days   (3) 34 days
(2) 27 days   (4) 41 days

62   When viewed from the planet, the moon has the greatest apparent diameter at point
(1) A (3) C
(2) B (4) D

63   If the distance from F_1 to F_2 is 42,000 kilometers and the distance from A to C is 768,000 kilometers, what is the eccentricity of the moon's orbit?
(1) 0.055   (3) 0.81
(2) 0.18   (4) 0.94

64   For an observer on the planet, at which position in the moon's orbit does the full-moon phase occur?
(1) A (3) C
(2) B (4) D

65   As the moon moves in its orbit from point D to point B, the force of gravitational attraction between the moon and the planet
1   increases, only
2   decreases, only
3   increases, then decreases
4   decreases, then increases

Group 3
If you choose this group, be sure to answer questions 66-70.

Base your answers to questions 66 through 70 on the *Earth Science Reference Tables*, the diagrams below, and your knowledge of Earth Science. The diagrams show the steps used to determine the amount of heat held by equal masses of iron, copper, lead, and granite.

66 This method of determining the amount of heat absorbed by substances assumes that the energy lost by a heat source is
1 refracted by a heat sink
2 reflected by a head sink
3 absorbed by a heat sink
4 scattered by a heat sink

STEP A
Measure the mass of each sample.

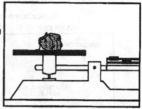

67 Which substance has the highest specific heat?
1 copper
2 granite
3 iron
4 lead

STEP B
Place samples in boiling water.

68 Why must the water be kept boiling in step *B*?
1 All samples must be heated to the same high temperature.
2 Boiling changes the melting point of the materials being tested.
3 The samples must be heated above 100°C.
4 Less energy is lost during a phase change.

STEP C
Measure the temperature of water in an insulated container.

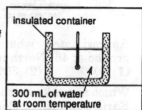

69 The granite sample is transferred from the boiling water to the room-temperature water. Why is the total heat lost by the granite greater than the total heat gained by the room-temperature water?
1 The granite sample had less volume than the other samples.
2 The granite sample lost some heat to the air as it was being transferred.
3 Water gained heat from the insulated container.
4 Water has a lower specific heat than the granite sample.

STEP D
Transfer one sample from the boiling water and stir gently with the thermometer. Record the temperature changes of the water.

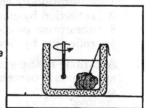

STEP E
Repeat steps C and D with each sample.

70 The movements of water molecules that transfer heat from one place to another within the water are called
1 radiation waves
2 transverse waves
3 conduction collisions
4 convection currents

Group 4
If you choose this group, be sure to answer questions 71-75.

Base your answers to questions 71 through 75 on the *Earth Science Reference Tables*, the diagrams and graph below, and your knowledge of Earth Science. The diagrams show the general effect of the Earth's atmosphere on insolation from the Sun at middle latitudes during both clear-sky and cloudy-sky conditions. The graph shows the percentage of insolation reflected by the Earth's surface at different latitudes in the Northern Hemisphere in winter.

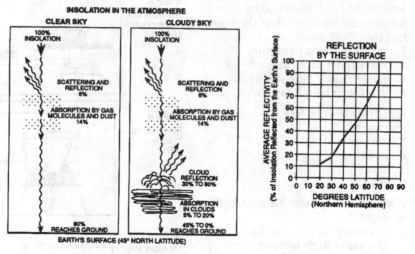

71 Approximately what percentage of the insolation actually reaches the ground at 45° North latitude on a clear day?
(1) 100%   (2) 80%   (3) 60%   (4) 45%

72 Which factor keeps the greatest percentage of insolation from reaching the Earth's surface on cloudy days?
1   absorption by cloud droplets
2   reflection by cloud droplets
3   absorption by clear-air gas molecules
4   reflection by clear-air gas molecules

73 According to the graph, on a winter day at 70° North latitude, what approximate percentage of the insolation is reflected by the Earth's surface?
(1) 50%   (2) 65%   (3) 85%   (4) 45%

74 Which statement best explains why, at high latitudes, reflectivity of insolation is greater in winter than in summer?
1   The North Pole is tilted toward the Sun in winter.
2   Snow and ice reflect almost all insolation.
3   The colder air holds much more moisture.
4   Dust settles quickly in cold air.

75 The radiation that passes through the atmosphere and reaches the Earth's surface has the greatest intensity in the form of
1   visible-light radiation   3   ultraviolet radiation
2   infrared radiation   4   radio-wave radiation

Group 5
If you choose this group, be sure to answer questions 76-80.

Base your answers to
questions 76 through 80 on the
*Earth Science Reference
Tables*, the weather map at the
right showing part of the
United States, and your
knowledge of Earth Science.
Letters *A* through *E* represent
weather stations.

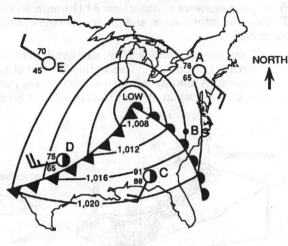

76   At which weather station
     is the barometric pressure
     reading most likely to be
     1,018.0 millibars?
     (1) *A*
     (2) *C*
     (3) *B*
     (4) *D*

77   Which weather station model best represents weather conditions at station
     B?

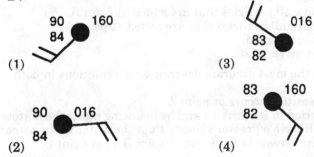

78   Which station's weather has been influenced most recently by the passage
     of a cold front?
     (1) *E*   (3) *C*
     (2) *B*   (4) *D*

79   At which weather station is precipitation most likely occurring at the
     present time?
     (1) *A*   (3) *E*
     (2) *B*   (4) *D*

80   If the low-pressure center follows a normal storm track, it will move toward
     the
     1   southeast   3   northeast
     2   southwest   4   northwest

Group 6
If you choose this group, be sure to answer questions 81-85.

Base your answers to questions 81 through 85 on the *Earth Science Reference Tables*, the information and diagrams below, and your knowledge of Earth Science.

> A mixture of colloids, clay, silt, sand, pebbles, and cobbles is put into stream I at point *A*. The water velocity at point *A* is 400 centimeters per second. A similar mixture of particles is put into stream II at point *A*. The water velocity in stream II at point A is 80 centimeters per second.

STREAM I   STREAM II

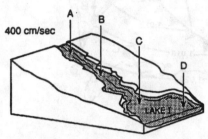
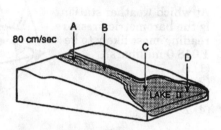

81   Which statement best describes what happens when the particles are placed in the streams?
     1   Stream I will move all particles that are added at point *A*.
     2   Stream II will move all particles that are added at point *A*.
     3   Stream I cannot move sand.
     4   Stream II cannot move sand.

82   Which statement is the most accurate description of conditions in both streams?
     1   The greatest deposition occurs at point *B*.
     2   Particles are carried in suspension and by bouncing along the bottom.
     3   The particles will have a greater velocity than the water in the stream.
     4   The velocity of the stream is the same at point *B* as at point *C*.

83   If a sudden rainstorm occurs at both streams above point *A*, the erosion rate will
     1   increase for stream I, but not for stream II
     2   increase for stream II, but not for stream I
     3   increase for both streams
     4   not change for either stream

84   What will most likely occur when the transported sediment reaches lake II?
     1   Clay particles will settle first.
     2   The largest particles will be carried farthest into the lake.
     3   The sediment will become more angular because of abrasion.
     4   The particles will be deposited in sorted layers.

Note that question 85 has only three choices.
85   In lake I, as the stream water moves from point C to point D, its velocity
     1   decreases
     2   increases
     3   remains the same

Group 7
If you choose this group, be sure to answer question 86-90.

Base your answers to questions 86 through 90 on the *Earth Science Reference Tables*, the diagram below, and your knowledge of Earth Science. The diagram represents three cross sections of the Earth at different locations a depth of 50 kilometers below sea level. The measurements given with each cross section indicate the thickness and density of the layers.

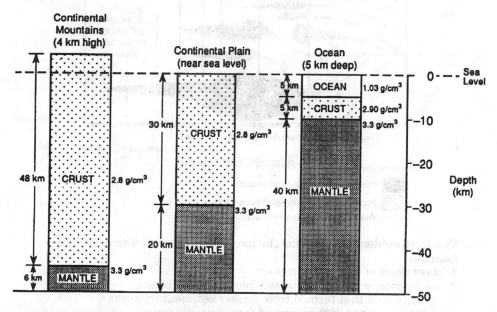

86 In which group are the layers of the Earth arranged in order of increasing average density?
1   mantle, crust, ocean water   3   ocean water, mantle, crust
2   crust, mantle, ocean water   4   ocean water, crust, mantle

87 Which material is most likely to be found 20 kilometers below sea level at the continental mountain location?
1   basalt   3   shale
2   granite   4   limestone

88 Which statement about the Earth's mantle is confirmed by the diagram?
1   The mantle is liquid.
2   The mantle has the same composition as the crust.
3   The mantle is located at different depths below the Earth's surface.
4   The mantle does not exist under continental mountains.

89 Compared with the oceanic crust, the continental crust is
1   thinner and less dense   3   thicker and less dense
2   thinner and more dense   4   thicker and more dense

90 The division of the Earth's interior into crust and mantle, as shown in the diagram, is based primarily on the study of
1   radioactive dating   3   volcanic eruption
2   seismic waves   4   gravity measurements

Group 8
If you choose this group, be sure to answer questions 91-95.

Base your answers to questions 91 through 95 on the *Earth Science Reference Tables*, the graph below, and your knowledge of Earth Science. The graph shows the development, growth in population, and extinction of the six major groups of trilobites, labeled *A* through *F*.

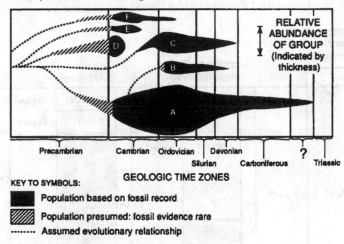

RELATIVE
ABUNDANCE
OF GROUP
(Indicated by thickness)

GEOLOGIC TIME ZONES

KEY TO SYMBOLS:

Population based on fossil record

Population presumed: fossil evidence rare

-------- Assumed evolutionary relationship

91 The fossil evidence that forms the basis for this graph was most likely found in
  1 lava flows of ancient volcanoes
  2 sedimentary rock that formed from ocean sediment
  3 granite rock that formed from former sedimentary rocks
  4 metamorphic rock that formed from volcanic rocks

92 Which group of trilobites became the most abundant?
  (1) *A*   (2) *B*   (3) *C*   (4) *D*

93 During which period did the last of these trilobite groups become extinct?
  1 Cretaceous   3 Permian
  2 Triassic   4 Carboniferous

94 Which inference is best supported by the graph?
  1 All trilobites evolved from group *A* trilobites.
  2 Trilobite groups became most abundant during the Devonian Period.
  3 Precambrian trilobite fossils are very rare.
  4 Trilobites could exist in present-day marine climates.

95 The diagrams below represent rock outcrops in which the rock layers have not been overturned. Which rock outcrop shows a possible sequence of the trilobite fossils?

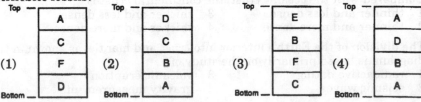

Group 9
If you choose this group be sure to answer questions 96-100.

Base your answers to questions 96 through 100 on the *Earth Science Reference Tables*, the diagrams below, and your knowledge of Earth Science. Diagram I shows a map of Niagara Falls and part of the Niagara River Gorge. The past locations of the edge of the river gorge are labeled in calendar years from 1678 to 1969. The lines show the changing locations of the edge of the Horseshoe Falls as it receded (eroded upstream) due to weathering and undercutting. Diagram II shows a geologic cross section of the rock layers at the edge of the Niagara River Gorge.

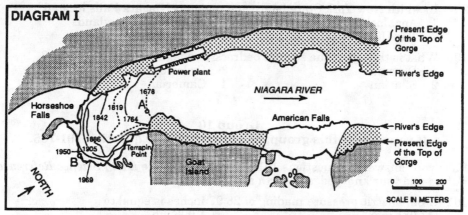

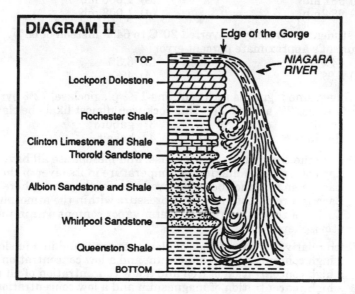

96 According to the map and scale, what is the approximate rate at which the Horseshoe Falls receded upstream from point *A* in 1678 to point *B* in 1969?
    (1) 1 meter per year
    (2) 0.1 meter per year
    (3) 10 meters per year
    (4) 200 meters per year

97   According to the geologic cross section (diagram II), the top rock layer at Niagara Falls consists of
1   marine-derived sediments compacted and cemented together
2   crystalline minerals resulting from melting and solidification
3   banded and distorted layers due to extreme pressure
4   metamorphosed clay, sand, and volcanic ash

98   Most scientists believe that the Niagara River started flowing at the end of the last ice age. During which geologic epoch did this ice age occur?
1   Early Triassic   3   Miocene
2   Paleocene   4   Pleistocene

99   In which New York State landscape region is Niagara Falls located?
1   Appalachian Plateau   3   Erie-Ontario Lowlands
2   Tug Hill Plateau   4   St. Lawrence Lowlands

100   What is the age of the surface bedrock at Niagara Falls, New York?
1   Devonian   3   Ordovician
2   Silurian   4   Cambrian

Group 10
If you choose this group, be sure to answer question 101-105.

Base your answers to questions 101 through 105 on the *Earth Science Reference Tables* and your knowledge of earth science.

101   A barometric pressure reading of 28.97 inches is equal to
(1) 981 mb   (3) 1,006 mb
(2) 984 mb   (4) 1,008 mb

102   A student incorrectly converted 20°C to 64°F instead of 68°F. What is the student's approximate percent error?
(1) 44%   (3) 6.3%
(2) 5.9%   (4) 4%

103   A fine-grained igneous rock contains 11% plagioclase, 72% pyroxene, 15% olivine, and 2% amphibole. This rock would most likely be classified as
1   granite   3   gabbro
2   rhyolite   4   basalt

104   What do the tropopause, stratopause, and mesopause all have in common?
1   Each is a point of maximum temperature in its layer of the atmosphere.
2   Each is an interface between two layers of the atmosphere.
3   Each is a region of increasing pressure within the atmosphere.
4   Each is a zone of decreasing water vapor content within the atmosphere.

105   Which relative concentrations of elements are found in a felsic rock?
1   a high concentration of aluminum and a low concentration of iron
2   a high concentration of iron and a low concentration of aluminum
3   a high concentration of magnesium and a low concentration of iron
4   a high concentration of magnesium and a low concentration of aluminum

Examination
Earth Science

	Raw Score Total on Part I ....
	Raw Score Total on Part II ....
	Raw Score Total on Lab Exam ....
	Total (Official Regents Examination Mark)

Pupil _____

School _____

Teacher_____

Part I (55 credits)

1	1 2 3 4	21	1 2 3 4	41	1 2 3 4
2	1 2 3 4	22	1 2 3 4	42	1 2 3 4
3	1 2 3 4	23	1 2 3 4	43	1 2 3 4
4	1 2 3 4	24	1 2 3 4	44	1 2 3 4
5	1 2 3 4	25	1 2 3 4	45	1 2 3 4
6	1 2 3 4	26	1 2 3 4	46	1 2 3 4
7	1 2 3 4	27	1 2 3 4	47	1 2 3 4
8	1 2 3 4	28	1 2 3 4	48	1 2 3 4
9	1 2 3 4	29	1 2 3 4	49	1 2 3 4
10	1 2 3 4	30	1 2 3 4	50	1 2 3 4
11	1 2 3 4	31	1 2 3 4	51	1 2 3 4
12	1 2 3 4	32	1 2 3 4	52	1 2 3 4
13	1 2 3 4	33	1 2 3 4	53	1 2 3 4
14	1 2 3 4	34	1 2 3 4	54	1 2 3 4
15	1 2 3 4	35	1 2 3 4	55	1 2 3 4
16	1 2 3 4	36	1 2 3 4		
17	1 2 3 4	37	1 2 3 4		
18	1 2 3 4	38	1 2 3 4		
19	1 2 3 4	39	1 2 3 4		
20	1 2 3 4	40	1 2 3 4		

Part II (35 credits)
Answer the questions in only seven of the ten groups in this part.

Group 1	
56	1 2 3 4
57	1 2 3 4
58	1 2 3 4
59	1 2 3 4
60	1 2 3 4

Group 2	
61	1 2 3 4
62	1 2 3 4
63	1 2 3 4
64	1 2 3 4
65	1 2 3 4

Group 3	
66	1 2 3 4
67	1 2 3 4
68	1 2 3 4
69	1 2 3 4
70	1 2 3 4

Group 4	
71	1 2 3 4
72	1 2 3 4
73	1 2 3 4
74	1 2 3 4
75	1 2 3 4

Group 5	
76	1 2 3 4
77	1 2 3 4
78	1 2 3 4
79	1 2 3 4
80	1 2 3 4

Group 6	
81	1 2 3 4
82	1 2 3 4
83	1 2 3 4
84	1 2 3 4
85	1 2 3 4

Group 7	
86	1 2 3 4
87	1 2 3 4
88	1 2 3 4
89	1 2 3 4
90	1 2 3 4

Group 8	
91	1 2 3 4
92	1 2 3 4
93	1 2 3 4
94	1 2 3 4
95	1 2 3 4

Group 9	
96	1 2 3 4
97	1 2 3 4
98	1 2 3 4
99	1 2 3 4
100	1 2 3 4

Group 10	
101	1 2 3 4
102	1 2 3 4
103	1 2 3 4
104	1 2 3 4
105	1 2 3 4

Earth Science Practice Examination 2 - January 1994

Part I Answer all 55 questions in this part. (55)

Directions (1-55): For *each* statement or question, select the word or expression that, of those given, best completes the statement or answers the question.

page 263

1   The Earth's planetary winds are deflected as a result of the Earth's
    1   revolution around the Sun   3   rotation on its axis
    2   seasonal changes   4   tilted axis

2   Which statement best describes the major heat flow associated with an iceberg as it drifts south from the Arctic Ocean into warmer weather?
    1   Heat flows from the water into the ice.
    2   Heat flows from the ice into water.
    3   A state of equilibrium exists, with neither ice nor water gaining or losing energy.
    4   Heat flows equally from the ice and the water into the surrounding air.

3   According to the *Earth Science Reference Tables*, what is the approximate thickness of the troposphere?
    (1) 7 km   (2) 12 km   (3) 27 km   (4) 50 km

4   The actual polar diameter of the Earth is 12,714 kilometers. The equatorial diameter of the Earth is approximately
    (1) 12,671 km   (3) 12,714 km
    (2) 12,700 km   (4) 12,757 km

5   The phases of the Moon are caused by the
    1   rotation of the Earth on its axis
    2   rotation of the Moon on its axis
    3   revolution of the Moon around the Earth
    4   revolution of the Earth around the Sun

6   Which graph best shows the amount of the lighted part of the Moon that an observer on the Earth would see during 1 month, beginning with the new moon phase?

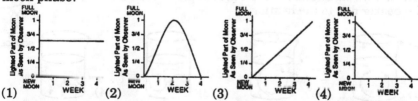

7   Which diagram best shows the altitude and direction of Polaris for an observer in New York City? [Refer to the *Earth Science Reference Tables.*]

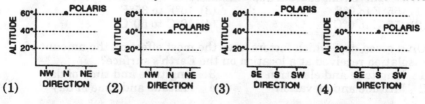

8   The diagram at the right shows the
    rotating Earth as it would appear from a
    satellite over the North Pole.

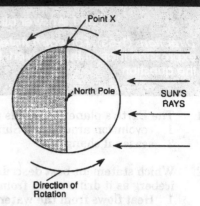

    The time at point *X* is closest to
    (1) 6 a.m.
    (2) 12 noon
    (3) 6 p.m.
    (4) 12 midnight

9   Equal masses of basalt, copper, granite, and iron that are at room
    temperature are places in boiling water. Which sample would reach the
    same temperature as the boiling water *first*? [Refer to the *Earth Science
    Reference Tables*.]
    1   basalt   2   copper   3   granite   4   iron

10 When a strong wind is blowing from one location to another, the two
    locations most likely have a difference in
    1   elevation   3   dewpoint temperature
    2   cloud cover   4   air pressure

11 When visible light strikes a snow-covered, flat field at a low angle, most of
    the energy will be
    1   absorbed by the snow   3   reflected by the snow
    2   refracted by the snow   4   radiated by the snow

12 On which date does maximum isolation usually occur in New York State?
    1   June 21   3   August 21
    2   July 10   4   August 31

13 Water is being heated in a beaker as shown at the right.

    Which drawing shows the most probable movement of water in
    the beaker due to the heating?

    (1)   (2)   (3)   (4)

14 According to the *Earth Science Reference Tables*, the temperature in the
    stratosphere ranges from approximately
    (1) -55°F to 0°F   (3) 10°F to 35°F
    (2) -55°C to 0°C   (4) 10°C to 50°C

15 On a given day, which factors have the most effect on the amount of
    insolation received at a location on the Earth's surface?
    1   longitude and elevation   3   longitude and time of day
    2   latitude and elevation   4   latitude and time of day

16 A cold front is moving eastward across New York State at an average speed of 50 kilometers per hour. Approximately how long will the front take to move from Buffalo to Albany? [Refer to the *Earth Science Reference Tables*.]
(1) 5 hours   (2) 8 hours   (3) 3 hours   (4) 10 hours

17 Atmospheric tranparency is most likely to increase after
   1   volcanic eruptions   3   industrial activity
   2   forest fires   4   precipitation

18 In the closed aquarium shown in the diagram at the right, the amount of water evaporating is equal to the amount of water vapor condensing.

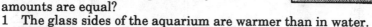

Which statement best explains why these amounts are equal?
   1   The glass sides of the aquarium are warmer than in water.
   2   The air in the aquarium is 50% saturated.
   3   The relative humidity outside the aquarium is 100%.
   4   The air in the aquarium is saturated.

19 Water vapor enters the atmosphere by the processes of evaporation and
   1   condensation   3   transpiration
   2   precipitation   4   conduction

20 Under which conditions would the volume of water in a stream most likely be greatest?
   1   Potential evapotranspiration is less than precipitation, and stored soil moisture is at capacity.
   2   Potential evapotranspiration is less than precipitation, and stored soil moisture is depleted.
   3   Potential evapotranspiration is greater than precipitation, and stored soil moisture is depleted.
   4   Potential evapotranspiration is greater than precipitation, and stored soil moisture is at capacity.

Base your answers to questions 21 and 22 on the isotherm map of North America and part of South America. The map shows the average daily temperature in degrees Fahrenheit during 1 month of the year.

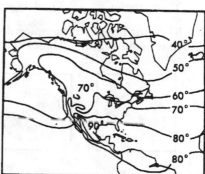

21 This map could represent the average daily temperature for the month of
   1   November
   2   January
   3   March
   4   July

22 Why does the 60 degree isotherm blend northward in the Northern Hemisphere during the time of year when the data was recorded?
   1   The land is warmer than the ocean.
   2   Warm ocean currents are moving northward along both coasts.
   3   The mid-ocean ridges are heating the ocean water.
   4   A high-pressure air mass is centered over North America.

23 During winter, New York City frequently receives rain when locations just
     north and west of the city receive snow. Which statement best explains
     this difference?
     1   The snow in the clouds has been depleted by the time the storm reaches
        New York City.
     2   The ocean modifies New York City's temperatures.
     3   New York City usually receives its weather from the south.
     4   New York City has a higher elevation.

24 Which graph best illustrates the relationship between the slope of the land
     and the amount of surface runoff?

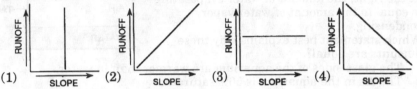

25 Which statement best explains why a
     desert often forms on the leeward side
     of a mountain range, as shown in the
     diagram at the right?
     1   Sinking air compresses and warms.
     2   Sinking air expands and warms.
     3   Rising air compresses and warms.
     4   Rising air expands and warms.

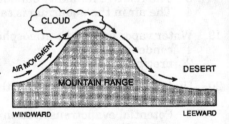

26 In summer, a small stream has a depth of 3 meters and a velocity of 0.5
     meter per second. In spring, the same stream has a depth of 5 meters. The
     velocity of the stream in spring is most likely closest to
     (1) 0.1 m/sec   (2) 0.2 m/sec   (3) 0.5 m/sec   (4) 0.8 m/sec

27 Which substance has the greatest effect on the rate of weathering of rock?
     1   nitrogen   2   hydrogen   3   water   4   argon

28 The four limestone samples illustrated below have the same composition,
     mass, and volume. Under the same climatic conditions, which sample will
     weather fastest?

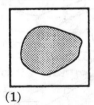 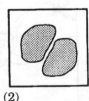 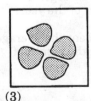 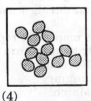

(1)   (2)   (3)   (4)

29 According to the Rock Cycle diagram in the *Earth Science Reference
     Tables,* which type(s) of rock can be the source of deposited sediments?
     1   igneous and metamorphic rocks, only
     2   metamorphic and sedimentary rocks, only
     3   sedimentary rocks, only
     4   igneous, metamorphic, and sedimentary rocks

30　The physical properties of a mineral sample are most closely related to the
　　1　arrangement of the mineral's atom
　　2　age of the mineral sample
　　3　size of the mineral sample
　　4　temperature of the mineral sample

31　According to the *Earth Science Reference Tables*, which sedimentary rock
　　most likely formed as an evaporite?
　　1　siltstone　　　　　　　　　　3　gypsum
　　2　conglomerate　　　　　　　　4　shale

32　Which characteristic of an igneous rock would provide the most information
　　about the environment in which the rock solidified?
　　1　color　　　　　　　　　　　　3　hardness
　　2　texture　　　　　　　　　　　4　streak

33　Four samples of aluminum, *A*, *B*, *C*, and *D*, have
　　identical volumes and densities, but different
　　shapes. Each piece is dropped into a long tube
　　filled with water. The time each sample takes to
　　settle to the bottom of the tube is shown in the
　　table at the right.

Sample	Time to Settle (sec)
A	2.5
B	3.7
C	4.0
D	5.2

Which diagram most likely represents the shape
of sample *A*?

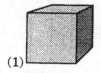 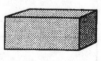 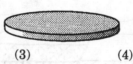

(1)　　　　　　　(2)　　　　　　(3)　　　　　　(4)

34　Where are the Earth's sedimentary rocks generally found?
　　1　in regions of recent volcanic activity
　　2　deep within the Earth's crust
　　3　along the mid-ocean ridges
　　4　as a thin layer covering much of the continents

35　Which combination of minerals would represent a rock categorized as being
　　in zone *C* of the Scheme for Igneous Rock Identification in the *Earth
　　Science Reference Tables*?
　　(1) 10% orthoclase, 30% quartz, 30% plagioclase, 15% mica, 15% amphibole
　　(2) 40% orthoclase, 35% quartz, 12% plagioclase, 8% mica, 5% amphibole
　　(3) 80% orthoclase, 10% plagioclase, 5% mica, 5% amphibole
　　(4) 5% quartz, 55% plagioclase, 10% mica, 30% amphibole

36　According to the *Earth Science Reference Tables*, how many million years
　　ago did the surface bedrock under Watertown, New York, form?
　　(1) 345 to 395　　　　　　　　(3) 435 to 500
　　(2) 395 to 435　　　　　　　　(4) 500 to 570

37   According to the *Earth Science Reference Tables*, which radioactive element would be most useful in determining the age of clothing that is thought to have been worn 2,000 years ago?
   1   carbon-14
   2   potassium-40
   3   uranium-238
   4   rubidium-87

38   The diagram at the right represents a clock used to time the half-life of a particular radioactive substance. The clock was started at 12:00. The shaded portion on the clock represents the number of hours one-half of this radioactive substance took to disintegrate.

   Which diagram best represents the clock at the end of the next half-life of this radioactive substance?

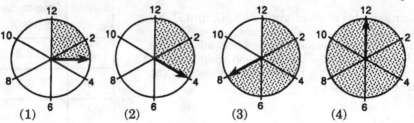

(1)   (2)   (3)   (4)

Base your answers to questions 39 and 40 on the geologic cross section below.

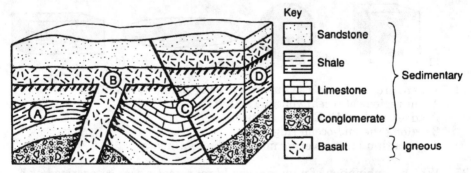

39   Which geologic event occurred most recently?
   1   folding at *A*
   2   the intrusion at *B*
   3   faulting at *C*
   4   the unconformity at *D*

40   The symbol ⁊⁊⁊⁊⁊⁊ in the diagram most likely represents a
   1   metamorphic rock in contact with an igneous rock
   2   depression caused by underground erosion
   3   convection cell caused by unequal heating
   4   large fault from an earthquake or crustal movement

41 The circles on the map at the right show the distances from three seismic stations, *X*, *Y*, and *Z*, to the epicenter of an earthquake.

Which location is closest to the earthquake epicenter?
(1) *A*
(2) *B*
(3) *C*
(4) *D*

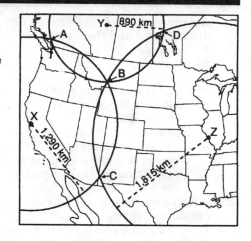

42 According to the *Earth Science Reference Tables*, during which geologic period were the continents all part of one landmass, with North America and South America joined to Africa?
1 Tertiary   3 Triassic
2 Cretaceous   4 Carboniferous

43 Which statement correctly describes the Earth's crust?
1 It is of uniform thickness.
2 It is thicker under the poles than under the Equator.
3 It is thinnest under the center of continents.
4 It is thinnest under the oceans.

44 A seismic station is 2,000 kilometers from an earthquake epicenter. According to the *Earth Science Reference Tables*, how long does it take an *S*-wave to travel from the epicenter to the station?
(1) 7 minutes 20 seconds   (3) 3 minutes 20 seconds
(2) 5 minutes 10 seconds   (4) 4 minutes 10 seconds

45 The diagrams below represent profiles of four different landscapes, *A*, *B*, *C*, and *D*.

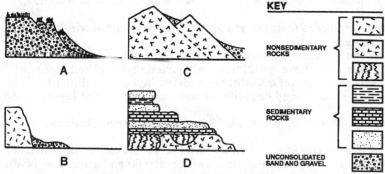

Which landscape is most likely to have a noticeable change in its profile after a heavy rainstorm?
(1) *A*   (2) *B*   (3) *C*   (4) *D*

46   According to the *Earth Science Reference Tables*, during which geologic era
       did trilobites and sharks coexist?
       1   Cenozoic   3   Paleozoic
       2   Mesozoic   4   Precambrian

47   Pleistocene deposits of gravel are found lying directly on Precambrian
       bedrock near Mt. Marcy, New York. The interface between the gravel and
       the bedrock indicates
       1   a zone of contact metamorphism
       2   an area of volcanic activity that resulted in extruded gravels
       3   a period of continuous deposition
       4   a major time gap in the geologic record

48   Which change is most likely to occur in a landscape region that is uplifted
       rapidly by folding?
       1   The climate will become warmer.
       2   The stream drainage patterns will change.
       3   The composition of the bedrock will change.
       4   The hillslopes will become less steep.

49   According to the *Earth Science Reference Tables*, which event accounted for
       the vast surface changes that occurred in New York State during the
       Pleistocene Epoch?
       1   the uplift of the Adirondack Mountains
       2   the shifting of the North American continent toward the Equator
       3   the intrusion of the Palisades sill
       4   the advance and retreat of the last continental ice sheet

50   According to the *Earth Science Reference Tables*, the Appalachian Uplands
       region of New York State is classified as which type of landscape?
       1   plateau   3   lowland
       2   highland   4   coastal plain

51   According to the *Earth Science Reference Tables*, most of the metamorphic
       surface bedrock in New York State is located in which landscape region?
       1   Atlantic Coastal Lowlands   3   Adirondack Highlands
       2   Appalachian Uplands   4   Erie-Ontario Lowlands

52   Which landscape features are primarily the result of wind erosion and
       deposition?
       (1) U-shaped valleys containing unsorted layers of sediment
       (2) V-shaped valleys containing well-sorted layers of sediment
       (3) terraces of gravel containing unsorted layers of sediment
       (4) cross-bedded sand deposits containing finely sorted layers of sediment

Note that questions 53 through 55 have only three choices.

53   As the quantities of water vapor and carbon dioxide in the Earth's
       atmosphere increase, the amount of terrestrial radiation that is absorbed
       by the atmosphere normally
       1   decreases
       2   increases
       3   remains the same

54 According to the *Earth Science Reference Tables,* as the dewpoint
    temperature of a sample of air decreases, the amount of moisture in that
    sample of air
    1 decreases   2 increases   3 remains the same

55 Compared to the velocity of an earthquake's *P*-waves, the velocity of the
    S-waves in the same material is
    1 less   2 greater   3 the same

Part II

**This part consists of ten groups, each containing five questions. Choose
seven of these ten groups. Be sure that you answer all five questions in
each group chosen. Record the answers to these questions on the
seperate answer sheet in accordance with the directions on the front
page of this booklet.** [35]

Group 1
If you choose this group, be sure to answer questions 56-60.

Base your answers to
questions 56 through 60 on
the data table at the right
and your knowledge of
Earth Science. The data
table shows the air pressure
and air temperatures
collected by nine observers
at different elevations on the
same side of a high
mountain. The data were
collected at 12:00 noon on a
clear, calm day.

Station	Elevation (m)	Air Pressure (mb)	Air Temperature (°C)
1	sea level	1,000	22
2	200	980	20
3	400	960	18
4	600	940	16
5	800	920	14
6	1,000	900	12
7	1,200	880	10
8	1,400	860	9
9	1,600	840	8

56 From sea level to an elevation of 1,200 meters, air pressure decreased at the
    rate of
    (1) 1.0 mb/m   (3) 10.0 mb/m
    (2) 0.1 mb/m   (4) 100.0 mb/m

57 Based on the data collected, which graph best represents the relationship
    between elevation above sea level and air pressure?

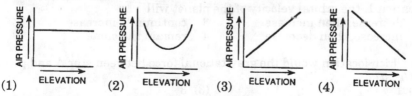

58 Which station model best represents weather conditions at station 1at the
    same time?

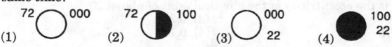

59   The change in temperature between station 1 and station 9 is most likely
      due to modification of temperature patterns by
      1   planetary wind belts   3   ocean currents
      2   elevation above sea level   4   latitudes

60   At 2:00 p.m. on the same day, the observers reported continuing clear and
      calm weather conditions. At that time, the air temperature at station 5
      would most likely have been
      (1) 9°C   (2) 11°C   (3) 13°C   (4) 15°C

Group 2
If you choose this group, be sure to answer questions 61-65.

Base your answers to
questions 61 through 65 on
the *Earth Science Reference
Tables*, the diagram at the
right, and your knowledge of
Earth Science. The diagram
represents planet Z in its
orbit around star A.
Locations 1 through 4 of
planet Z are indicated on the
orbit. The sizes of the planet
and the star are not drawn to
scale. The elliptical orbit of
planet Z and the distance
between the foci ($F1$ and $F2$)
are drawn to scale.

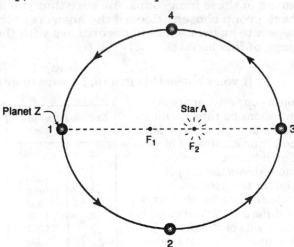

61   Star A would have the greatest apparent size to an observer on planet Z
      when the planet is at location
      (1) 1   (2) 2   (3) 3   (4) 4

62   The orbiting motion of planet Z around star A is known as
      1   rotation   3   declination
      2   inclination   4   revolution

63   As planet Z travels around star A in a complete orbit starting from
      location 1, the orbital velocity of the planet will
      1   decrease, then increase   3   continually increase
      2   increase, then decrease   4   remain the same

64   At which location would the gravitational force between star A and planet Z
      be *least*?
      (1) 1   (3) 3
      (2) 2   (4) 4

65   What is the eccentricity of the elliptical orbit of planet Z?
      (1) 1.0   (3) 0.20
      (2) 0.75   (4) 0.10

Group 3
If you choose this group, be sure to answer questions 66-70.

Base your answers to questions 66 through 70 on the *Earth Science Reference Tables*, the contour map at the right, and your knowledge of Earth Science. Points A through F represent locations on the map.

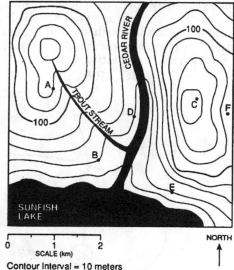

66 What is the most likely elevation of the surface of Sunfish Lake?
(1) 151 m
(2) 140 m
(3) 55 m
(4) 28 m

67 Which statement about hill C is best supported by the map?
1 Hill C is located approximately 2 km west of Cedar River.
2 The steepest slope of hill C is on the western side.
3 Hill C has been shaped by glaciers.
4 The highest possible elevation of hill C is 179 m.

68 If no elevation values were given, which general rule could be used to establish that Cedar River flows into Sunfish lake?
1 Rivers shown on maps generally flow southward.
2 Rivers always flow toward large bodies of water.
3 Contour lines bend upstream when crossing a river.
4 A large body of water is generally the source of water for a river.

69 Which location has the same elevation as location D?
(1) A   (2) E   (3) C   (4) F

70 Which diagram best represents the topographic profile from location A to location F?

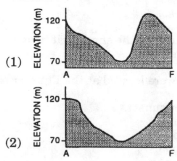

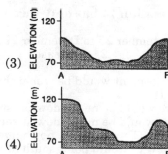

Group 4
If you choose this group, be sure to answer questions 71-75.

Base your answers to questions 71 through 75 on the *Earth Science Reference Tables,* the diagram and your knowledge of Earth Science. The diagram shows a post located in the Northern Hemisphere. Five different shadows, *A',B',C',D',* and *E',* are cast on a certain day by the post when the Sun is in positions *A,B,C,D,* and *E,* respectively.

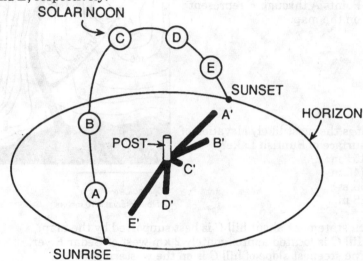

71　Which motion causes the apparent daily path of the Sun shown in the diagram?
1　the Sun's revolution　　　3　the Earth's revolution
2　the Sun's rotation　　　　4　the Earth's rotation

72　In the Northern Hemisphere, the intensity of insolation during the year is greatest when
1　shadow *A'* is longest　　　3　shadow *C'* is shortest
2　shadow *D'* is longest　　　4　shadow *E'* is shortest

73　What would be the approximate duration of insolation for this location when shadow *C'* reaches its greatest length during the year?
(1) 10 hours　　　　　　　(3) 15 hours
(2) 12 hours　　　　　　　(4) 24 hours

Note that question 74 has only three choices.

74　From September 23 to December 20, the length of the shadow at noon will
1　decrease　　　2　increase　　　3　remain the same

75　Which statement would be true if this post were located at the Equator on March 21?
1　There would be no shadows at sunrise or sunset.
2　There would be no shadow at solar noon.
3　Shadow *C'* would point north at solar noon.
4　Shadow *C'* would point south at solar noon.

Group 5
If you choose this group, be sure to answer questions 76-80.

Base your answers to questions 76 through 80 on the *Earth Science Reference Tables*, the diagram below, and your knowledge of Earth Science. The diagram represents s satellite image of Hurricane Gilbert in the Gulf of Mexico. Each **X** represents the position of the eye of the storm on the data indicated.

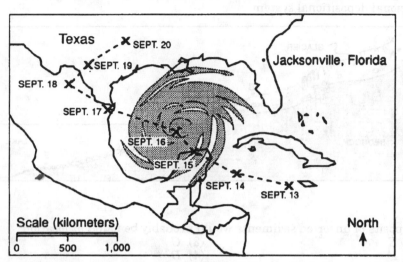

76 The general direction of Hurricane Gilbert's track from September 13 through September 18 was toward the
1   southwest   3   northwest
2   southeast   4   northeast

77 The surface wind pattern associated with Hurricane Gilbert was
1   counterclockwise and toward the center
2   counterclockwise and away from the center
3   clockwise and toward the center
4   clockwise and away from the center

78 What was the probable source of moisture for this hurricane?
1   carbon dioxide from the atmosphere
2   winds from the coastal deserts
3   transpiration from tropical jungles
4   evaporation from the ocean

79 On September 18, Hurricane Gilbert changed direction. Which statement provides the most probable reason for this change?
1   The air mass was cooled by the land surface.
2   The storm entered the prevailing westerlies wind belt.
3   The amount of precipitation released by the storm changed suddenly.
4   The amount of insolation received by the air mass decreased.

80 The air mass that gave rise to Hurricane Gilbert would be identified as
(1) cP   (2) cT   (3) mT   (4) mP

Group 6
If you choose this group, be sure to answer questions 81-85.

Base your answers to questions 81 through 85 on the *Earth Science Reference Tables*, the diagram below, and your knowledge of Earth Science. The diagram represents a glacier moving out of a mountain valley. The water from the melting glacier is flowing into a lake. Letters *A* through *F* identify points within the erosional/depositional system.

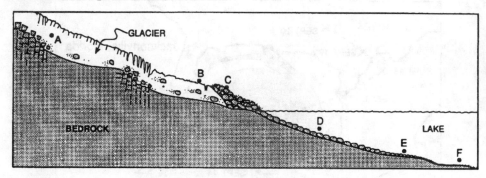

81   Deposits of unsorted sediments would probably be found at location
    (1) *E*   (3) *C*
    (2) *F*   (4) *D*

82   An interface between erosion and deposition by the ice is most likely located between points
    (1) *A* and *B*   (3) *C* and *D*
    (2) *B* and *C*   (4) *D* and *E*

83   Colloidal-sized sediment particles carried by water are most probably being deposited at point
    (1) *F*   (3) *C*
    (2) *B*   (4) *D*

84   Which characteristic would form as the glacier advances from point *A* to point *B*?
    (1) V-shaped valleys   (3) layers of salt and other evaporites
    (2) a thick, well-sorted soil   (4) scratched and polished bedrock

85   Which graph best represents the speed of a sediment particle as it moves from point *D* to point *F*?

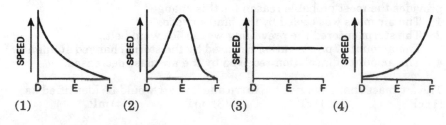

Group 7
If you choose this group, be sure to answer question 86-90.

Base your answers to questions 86 through 90 on the *Earth Science Reference Tables*, the data table below, and your knowledge of Earth Science. The table contains data taken at locations *A* through *E* in a stream. The volume of the stream is the same at all locations.

Location in the Stream	Average Velocity (cm/sec)	Elevation above Sea Level (m)	Distance from Source (km)
A	10	640	0
B	130	570	20
C	210	200	80
D	100	100	130
E	70	40	200

86 The velocity of the stream at a particular location is controlled mainly by the
1   elevation of the stream at the location
2   distance of the location from the source
3   slope of the streambed at the location
4   amount of sediments carried at the location

87 Which sediments transported by the stream at location *C* could *not* be transported by the stream at located *D*?
1   silt   3   sand
2   clay   4   cobbles

88 What is the gradient of the stream between locations *C* and *D*?
(1) 1.1 m/km   (3) 3.0 m/km
(2) 2.0 m/km   (4) 0.5 m/km

89 Sediments transported in the stream from location *B* to location C would most likely be transported by
1   suspension, only   3   solution and suspension, only
2   solution, only   4   solution, suspension, and rolling

90 Which diagram best represents the stream profile from location *A* to location *E*?

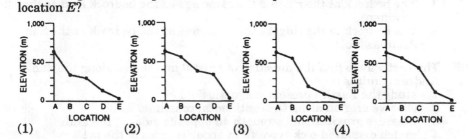

(1)   (2)   (3)   (4)

Group 8
If you choose this group, be sure to answer questions 91-95.

Base your answers to questions 91 through 95 on the map at the right and your knowledge of Earth Science. The map shows crustal plate boundaries located along the Pacific coastline of the United States. The arrows show the general directions in which some of the plates appear to be moving slowly.

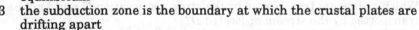

91 Which feature is located at 20° North latitude and 109° West longitude?
1   San Andreas Fault
3   Baja California
2   East Pacific rise
4   Juan de Fuca Ridge

92 Geologic studies of the San Andreas fault indicate that
1   many earthquakes occur along the San Andreas fault
2   the North American plate and the Pacific plate are locked in dynamic equilibrium
3   the subduction zone is the boundary at which the crustal plates are drifting apart
4   the age of the bedrock increases as distance from the fault increases

93 Which features are most often found at crustal plate boundaries like those shown on the map?
1   meandering rivers and warm-water lakes
2   plains and plateaus
3   geysers and glaciers
4   faulted bedrock and volcanoes

94 What would a study of the East Pacific rise (a mid-ocean ridge) indicate about the age of the basaltic bedrock in this area?
1   The bedrock is youngest at the ridge.
2   The bedrock is oldest at the ridge.
3   The bedrock at the ridge is the same age as the bedrock next to the continent.
4   The bedrock at the ridge is the same age as the bedrock at the San Andreas fault.

95 The best way to find the direction of crustal movement along the San Andreas fault is to
1   study the Earth's present magnetic field
2   observe erosion along the continental coastline
3   measure gravitational strength on opposite sides of the fault
4   match displaced rock types from opposite sides of the fault

Group 9
If you choose this group be sure to answer questions 96-100.

Base your answers to questions 96 through 100 on the *Earth Science Reference Tables*, the diagram below, and your knowledge of Earth Science. The diagram represents a geologic cross section of a portion of the Earth's crust. The rock layers have not been overturned.

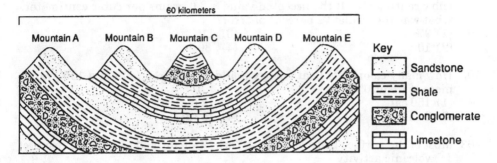

96 The top of which mountain is composed of the youngest bedrock?
    (1) *A*   (3) *C*
    (2) *B*   (4) *D*

97 What type(s) of bed rock can be found in this area?
    1   igneous, only
    2   sedimentary, only
    3   igneous and sedimentary, only
    4   igneous, sedimentary, and metamorphic

98 The two conglomerate layers represented in the diagram have the same texture, but only one layer contains sandstone pebbles. This observation leads to the inference that these two rock layers probably were
    1   affected by contact metamorphism
    2   solidified deep in the Earth's interior
    3   formed from sediments originating from different sources
    4   formed from sediments deposited at the same time

99 Which type of rock appears to be most resistant to weathering?
    1   sandstone   3   conglomerate
    2   shale   4   limestone

100 Which inference is best supported by this diagram?
    1   The region shows no evidence of crustal movement.
    2   The region shows evidence of several extinct volcanoes.
    3   The region has had extensive folding.
    4   The region has had extensive faulting.

Group 10
If you choose this group, be sure to answer question 101-105.

Base your answers to questions 101 through 105 on the *Earth Science Reference Tables* and your knowledge of Earth Science.

101 A student calculated the density of a mineral sample to be 2.7 grams per cubic centimeter. If the accepted value is 3.0 grams per cubic centimeter, what was the student's percent of error?
(1) 9%   (3) 11%
(2) 10%   (4) 30%

102 How many calories of heat energy would be required to melt a 150-gram piece of ice at 0°C?
(1) 150 cal   (3) 12,000 cal
(2) 2 cal   (4) 75 cal

103 Most of the surface bedrock of New York State formed as a direct result of
1 volcanic activity
2 spreading of the ocean floor
3 melting and solidification
4 compaction and cementation

104 The green sand found on some Hawaiian Island shorelines most probably consists primarily of
1 quartz   3 plagioclase feldspar
2 olivine   4 orthoclase feldspar

105 If a seismograph recording station located 5,700 kilometers from an epicenter receives a *P*-wave at 4:45 p.m., at which time did the earthquake actually occur at the epicenter?
(1) 4:24 p.m.   (3) 4:36 p.m.
(2) 4:29 p.m.   (4) 4:56 p.m.

Examination
Earth Science

Pupil _____

School _____

Teacher_____

Raw Score Total on Part I	
Raw Score Total on Part II	
Raw Score Total on Lab Exam	
Total (Official Regents Examination Mark)	

Part I (55 credits)

1	1 2 3 4	21	1 2 3 4	41	1 2 3 4
2	1 2 3 4	22	1 2 3 4	42	1 2 3 4
3	1 2 3 4	23	1 2 3 4	43	1 2 3 4
4	1 2 3 4	24	1 2 3 4	44	1 2 3 4
5	1 2 3 4	25	1 2 3 4	45	1 2 3 4
6	1 2 3 4	26	1 2 3 4	46	1 2 3 4
7	1 2 3 4	27	1 2 3 4	47	1 2 3 4
8	1 2 3 4	28	1 2 3 4	48	1 2 3 4
9	1 2 3 4	29	1 2 3 4	49	1 2 3 4
10	1 2 3 4	30	1 2 3 4	50	1 2 3 4
11	1 2 3 4	31	1 2 3 4	51	1 2 3 4
12	1 2 3 4	32	1 2 3 4	52	1 2 3 4
13	1 2 3 4	33	1 2 3 4	53	1 2 3 4
14	1 2 3 4	34	1 2 3 4	54	1 2 3 4
15	1 2 3 4	35	1 2 3 4	55	1 2 3 4
16	1 2 3 4	36	1 2 3 4		
17	1 2 3 4	37	1 2 3 4		
18	1 2 3 4	38	1 2 3 4		
19	1 2 3 4	39	1 2 3 4		
20	1 2 3 4	40	1 2 3 4		

Part II (35 credits)
Answer the questions in only seven of the ten groups in this part.

Group 1			
56	1 2 3 4
57	1 2 3 4
58	1 2 3 4
59	1 2 3 4
60	1 2 3 4

Group 2			
61	1 2 3 4
62	1 2 3 4
63	1 2 3 4
64	1 2 3 4
65	1 2 3 4

Group 3			
66	1 2 3 4
67	1 2 3 4
68	1 2 3 4
69	1 2 3 4
70	1 2 3 4

Group 4			
71	1 2 3 4
72	1 2 3 4
73	1 2 3 4
74	1 2 3 4
75	1 2 3 4

Group 5			
76	1 2 3 4
77	1 2 3 4
78	1 2 3 4
79	1 2 3 4
80	1 2 3 4

Group 6			
81	1 2 3 4
82	1 2 3 4
83	1 2 3 4
84	1 2 3 4
85	1 2 3 4

Group 7			
86	1 2 3 4
87	1 2 3 4
88	1 2 3 4
89	1 2 3 4
90	1 2 3 4

Group 8			
91	1 2 3 4
92	1 2 3 4
93	1 2 3 4
94	1 2 3 4
95	1 2 3 4

Group 9			
96	1 2 3 4
97	1 2 3 4
98	1 2 3 4
99	1 2 3 4
100	1 2 3 4

Group 10			
101	1 2 3 4
102	1 2 3 4
103	1 2 3 4
104	1 2 3 4
105	1 2 3 4

Earth Science Practice Examination 3 — June 1994

Part I Answer all 55 questions in this part. (55)

Directions (1-55): For *each* statement or question, select the word or expression that, of those given, best completes the statement or answers the question.

page 283

1   Which statement about a mineral sample found in a field in New York State is most likely an inference?
1   The sample was transported by a glacier.
2   The sample is white in color.
3   The sample is rectangular, with sharp, angular corners.
4   The sample is 8 cm long, 5 cm wide, and 3 cm high.

2   The use of a triple-beam balance to determine the mass of a rock is an example of measuring by using
1   all of the five senses
2   inferences and interpretations
3   a direct comparison with a standard
4   a combination of dimensional quantities

3   Which graph best represents the relationship between the latitude of an observer and the observed altitude of Polaris above the northern horizon?

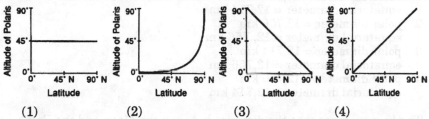

(1)   (2)   (3)   (4)

4   According to the *Earth Science Reference Tables*, as altitude increases from the tropopause to the mesopause, the atmospheric temperature will
1   decrease, only
2   increase, only
3   decrease, then increase
4   increase, then decrease

5   Which diagram best illustrates the heat transfer movement in fluids?

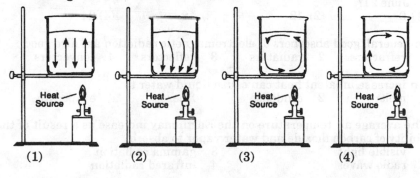

(1)   (2)   (3)   (4)

6   The diagrams below represent fossils found at different locations.

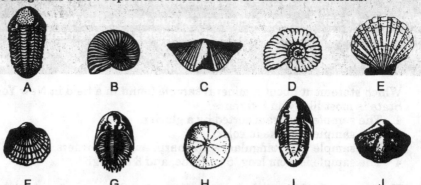

A   B   C   D   E

F   G   H   I   J

When classified by similarity of structure, which three fossils should be grouped together?
(1) *A, F,* and *H*   (3) *E, G,* and *H*
(2) *C, F,* and *J*   (4) *B, D,* and *I*

7   From which set of polar and equatorial diameters can the actual shape of the Earth be inferred?
1   polar diameter = 12,714 km
    equatorial diameter = 12,714 km
2   polar diameter = 12,756 km
    equatorial diameter = 12, 756 km
3   polar diameter = 12,714 km
    equatorial diameter = 12,756 km
4   polar diameter = 12,756 km
    equatorial diameter = 12,714 km

8   The best evidence that the distance between the Moon and the Earth varies is provided by the apparent change in the Moon's
1   shape   2   diameter   3   altitude   4   phase

9   Earth's atmospheric winds are deflected in a predictable manner because of the Earth's
1   rotation   2   revolution   3   gravity   4   inclination

10 Approximately how many hours of daylight are received at the North Pole on June 21?
(1) 0   (2) 12   (3) 18   (4) 24

11 In general, good absorbers of electromagnetic radiation are also good
1   refractors   2   radiators   3   reflectors   4   convectors

12 An increase in latent heat can cause liquid water to
1   melt   2   condense   3   freeze   4   evaporate

13 The average air temperature on the Earth may increase as a result of the ability of carbon dioxide and water vapor to absorb
1   visible light   3   gamma radiation
2   radio waves   4   infrared radiation

14 According to the *Earth Science Reference Tables*, an air pressure of 29.47
    inches of mercury is equal to
    (1) 996 mb
    (2) 998 mb
    (3) 1,002 mb
    (4) 1,014 mb

15 The graph below shows the changes in height of ocean water over the
    course of 2 days at one Earth location.

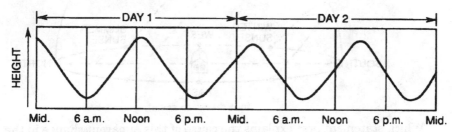

    Which statement concerning these changes is best supported by the graph?
    1   The changes are cyclic and occur at predictable time intervals.
    2   The changes are cyclic and occur at the same time every day.
    3   The changes are noncyclic and occur at sunrise and sunset.
    4   The changes are noncyclic and may occur at any time.

16 On which map of temperatures across the United States is the 60°F
    isotherm drawn correctly?

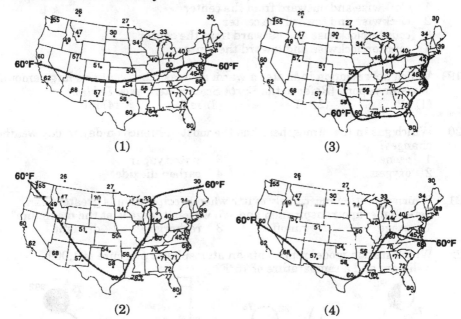

17   The diagram below shows the apparent paths of the Sun in relation to a
     house in New York State on June 21 and December 21.

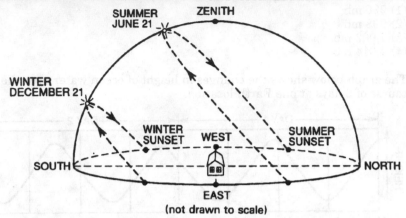

(not drawn to scale)

Which statement best explains the cause of this apparent change in the
Sun's path?
1   The Sun's orbital velocity changes as it revolves around the Earth.
2   The Earth's orbital velocity changes as it revolves around the Sun.
3   The Earth's axis is tilted 23½°.
4   The Sun's axis is tilted 23½°.

18   In the Northern Hemisphere, what is the direction of surface wind
     circulation is a high-pressure system?
1   clockwise and outward from the center
2   clockwise and toward the center
3   counterclockwise and outward from the center
4   counterclockwise and toward the center

19   Which abbreviation indicates a warm air mass that contains large amounts
     of water vapor? [Refer to the *Earth Science Reference Tables*.]
     (1) cP   (2) cT   (3) mT   (4) mP

20   Which gas in the atmosphere has the most influence on day-to-day weather
     changes?
1   ozone   3   water vapor
2   oxygen   4   carbon dioxide

21   Which condition most likely exists when precipitation is greater than
     potential evapotranspiration and soil water storage is at the maximum?
     1   usage   2   runoff   3   recharge   4   drought

22   Which station model represents an atmospheric pressure of 1,009.2
     millibars and a temperature of 75°F?

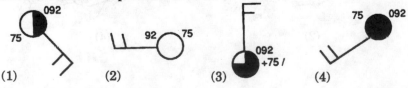

(1)   (2)   (3)   (4)

23   The diagram below represents snow falling on the Tug Hill Plateau in New York State.

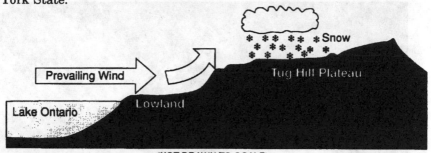

(NOT DRAWN TO SCALE)

The best explanation for the formation of snow under these conditions is that

1   dry air rises and warms   3   moist air rises and warms
2   moist air rises and cools   4   dry air sinks and cools

24   The diagram at the right represents the Earth's orbital path around the Sun. The Earth takes the same amount of time to move from *A* to *B* as from *C* to *D*.

Which values are equal within the system?

1   The shaded sections of the diagram are equal in area.
2   The distance from the Sun to the Earth is the same at point *A* and at point *D*.
3   The orbital velocity of the Earth at point *A* equals its orbital velocity at point *C*.
4   The gravitational force between the Earth and the Sun at point *B* is the same as the gravitational force at point *D*.

25   The graph below represents the average temperature of a city for each month of the year.

Where is the city most likely to be located?

1   inland in the Northern Hemisphere, in a middle latitude
2   inland in the Southern Hemisphere, in a middle latitude
3   on a coast near the Equator
4   on a coast in the Antarctic

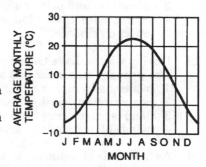

26   According to the *Earth Science Reference Tables*, which element is most abundant by mass in the Earth's crust?

1   nitrogen   3   silicon
2   oxygen   4   iron

27   The photograph below shows a fan-shaped accumulation of sediment.

This accumulation of sediment is the direct result of
1   weathering of bedrock   3   deposition by running water
2   erosion by wind   4   transport by glaciers

28   What is the best explanation for the two statements below?

- Some mountains located near the Earth's Equator have snow-covered peaks.
- Icecaps exist at the Earth's poles.

1   High elevation and high latitude have a similar effect on climate.
2   Both mountain and polar regions receive more energy from the Sun than other regions do.
3   Mountain and polar regions have arid climates.
4   An increase in snowfall and an increase in temperature have a similar effect on climate.

29   During winter, Lake Ontario is generally warmer than adjacent land areas. The primary reason for this temperature difference is that
1   water has a higher specific heat than land has
2   water reflects sunlight better than land does
3   land is more dense than water is
4   winds blow from land areas toward the water

30   As a particle of sediment in a stream breaks into several smaller pieces, the rate of weathering of the sediment will
1   decrease due to a decrease in surface area
2   decrease due to an increase in surface area
3   increase due to a decrease in surface area
4   increase due to an increase in surface area

31 When minerals are dissolved, how are the resulting ions carried by rivers?
    1   by precipitation   3   in suspension
    2   by tumbling and rolling   4   in solution

32 The relative hardness of a mineral can best be tested by
    1   scratching the mineral across a glass plate
    2   squeezing the mineral with calibrated pliers
    3   determining the density of the mineral
    4   breaking the mineral with a hammer

33 The diagrams below show how plant materials are changed into the three
    forms of coal by natural processes.

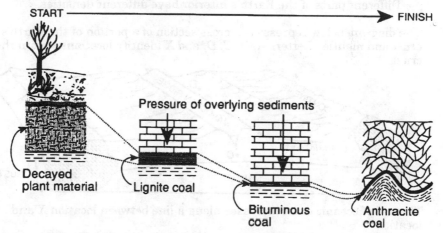

By which process is anthracite coal formed from bituminous coal? [Refer to
the *Earth Science Reference Tables*.]
    1   solidification   3   deposition
    2   metamorphism   4   intrusion

34 In the diagram at the right, the
    arrow shows the direction of
    stream flow around a bend.

    At which point does the greatest
    stream erosion occur?
    (1) *A*
    (2) *B*
    (3) *C*
    (4) *D*

35 Which is an accurate statement about rocks?
    1   Rocks are located only in continental areas of the Earth.
    2   Rocks seldom undergo change.
    3   Most rocks contain fossils.
    4   Most rocks have several minerals in common.

36   The size of the mineral crystals found in an igneous rock is directly related to the
     1   density of the minerals
     2   color of the minerals
     3   cooling time of the molten rock
     4   amount of sediments cemented together

37   Which statement best explains why the direction of some seismic waves changes sharply as the waves travel through the Earth?
     1   The Earth is spherical.
     2   Seismic waves tend to travel in curved paths.
     3   The temperature of the Earth's interior decreases with depth.
     4   Different parts of the Earth's interior have different densities.

38   The diagram below represents a cross section of a portion of the Earth's crust and mantle. Letters A, B, C, D, and X identify locations within the crust.

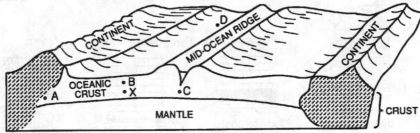

The age of oceanic crust increases along a line between location X and location
(1) A (2) B (3) C (4) D

39   In a soil sample, the particles have the same shape but different sizes. Which graph best represents the relationship between particle size and settling time when these particles are deposited in a quiet body of water?

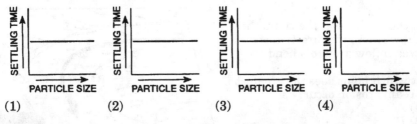

(1)   (2)   (3)   (4)

40   The best evidence of crustal uplift would be provided by
     1   marine fossils in the Rocky Mountains
     2   sediments in the Gulf of Mexico
     3   trenches in the Pacific Ocean floor
     4   igneous rock deep within the Earth

41   According to the *Earth Science Reference Tables*, what is the approximate total distance traveled by an earthquake's P-wave in its first 9 minutes?
     (1) 2,600 km   (2) 5,600 km   (3) 7,600 km   (4) 12,100 km

42 Which evidence supports the theory of ocean-floor spreading?
1   The rocks of the floor and the continents have similar origins.
2   In the ocean floor, rocks near the mid-ocean ridge are cooler than rocks near the continents.
3   The pattern of magnetic orientation of rocks is similar on both sides of the mid-ocean ridge.
4   The density of oceanic crust is greater than the density of continental crust.

43 A layer of volcanic ash may serve as a time marker because the ash is
1   generally deposited only on land
2   composed of index fossils
3   deposited rapidly over a large area
4   often a distinct color

44 One similarity between uranium-238 and carbon-14 is that both
1   decay at a predictable rate
2   have the same half-life
3   are normally found in large quantities in living matter
4   are found in granite

45 The diagram below represents a series of brachiopod fossils showing progressive changes during the Early Mississippian Epoch. The fossils are drawn to scale.

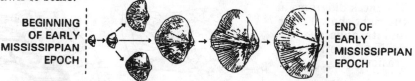

One explanation for this process of change is the theory of
1   superposition   3   dynamic equilibrium
2   evolution   4   fossilization

46 What is the best interpretation of the two statements below?
    • Corals are marine animals that live in warm ocean water.
    • Fossil corals are found in surface bedrock in areas of NYS.
1   Corals once lived on land.
2   Corals have migrated northward.
3   Parts of New York State are now covered by warm ocean water.
4   Parts of New York State were once covered by warm ocean water.

47 Which feature often indicates a boundary between landscape regions?
1   a highway cutting through a mountain region
2   resistant bedrock composed of more than one type of mineral
3   a long, meandering stream flowing across a large, level region
4   a change in slope between adjoining bedrock types with different structures

48 According to the *Earth Science Reference Tables*, when did the intrusion of the Palisades Sill occur?
1   before the Appalachian Orogeny
2   during the late Triassic Period
3   during the Paleozoic Era
4   after the extinction of dinosaurs and ammonites

49 According to the *Earth Science Reference Tables*, what are the respective decay products of uranium, potassium, and rubidium?
1 lead (Pb), argon (Ar), and strontium (Sr)
2 carbon (C), oxygen(O), and nitrogen (N)
3 hydrogen (H), lithium (Li), and helium (He)
4 silicon (Si), oxygen (O), and aluminum (Al)

50 According to the *Earth Science Reference Tables*, at which location could a geologist find shale containing eurypterid fossils?
1 Old Forge   3 New York City
2 Syracuse   4 Long Island

51 The advance and retreat of continental ice sheets produced deposits of sands and gravels during the Pleistocene Epoch. According to the *Earth Science Reference Tables*, in which New York State landscape region were such sands and gravels deposited over Cretaceous and Tertiary materials?
1 Atlantic Coastal Plain   3 Adirondack Mountains
2 Erie-Ontario Lowlands   4 Tug Hill Plateau

52 According to the *Earth Science Reference Tables*, in which type of landscape region in Elmira, New York, located?
1 mountains   2 plain   3 plateau   4 lowlands

53 The block diagrams at the right show a river and its landscape during four stages of erosion.

In which order should the diagrams be placed to show the most likely sequence of river and landscape development?
(1) *A, D, B, C*
(2) *B, D, C, A*
(3) *C, B, A, D*
(4) *D, A, C, B*

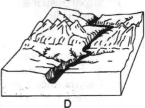
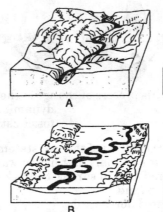

54 Which condition is a characteristic of a landscape region in dynamic equilibrium?
1 a balance uplifting and erosion
2 a balance weathering and erosion
3 more erosion than uplifting
4 more uplifting than weathering

Note that question 55 has only three choices.

55 A person in New York State observes a star that is due east and just above the horizon. During the next hour, the distance between the star and the horizon will appear to
1 decrease   2 increase   3 remain the same

Part II

This part consists of ten groups, each containing five questions. Choose seven of these ten groups. Be sure that you answer all five questions in each group chosen. Record the answers to these questions on the separate answer sheet in accordance with the directions on the front page of this booklet.   [35]

Group 1
If you choose this group, be sure to answer questions 56-60.

Base your answers to questions 56 through 60 on the *Earth Science Reference Tables,* the contour map below, and your knowledge of Earth Science. Letters *A* through *K* represent locations in the area. Hachure lines (⊥⊥⊥⊥) show depressions.

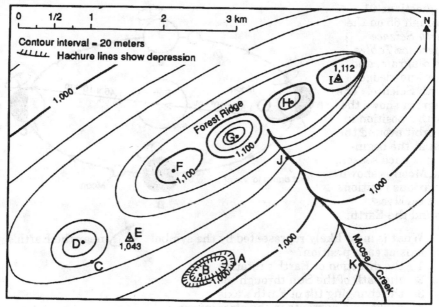

56   Which hilltop could have an elevation of 1,145 meters?
  (1) *D*   (2) *F*   (3) *G*   (4) *H*

57   Toward which direction does Moose Creek flow?
  1   southeast   3   southwest
  2   northeast   4   northwest

58   Which graph best represents the map profile along a straight line from point *C* through point *A* to point *K*?

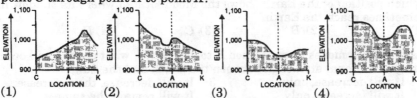

59 Which equation would be used to determine the stream gradient along Moose Creek between points J and K?

1 gradient $= \dfrac{1.8 \text{ km}}{80 \text{ m}}$ x 100 3 gradient $= (1{,}040 \text{ m} - 960 \text{ m}) \times 20 \text{ m}$

2 gradient $= \dfrac{0.8 \text{ km}}{60 \text{ m}}$ 4 gradient $= \dfrac{80 \text{ m}}{1.8 \text{ km}}$

60 What is the *lowest* possible elevation of point B?
(1) 981 m (2) 971 m (3) 961 m (4) 941 m

Group 2
If you choose this group, be sure to answer questions 61-65.

Base your answers to questions 61 through 65 on the *Earth Science Reference Tables*, the diagram, and your knowledge of Earth Science. The diagram shows the Earth's position in its orbit around the Sun at the beginning of each season. The Moon is shown at various positions as it revolves around the Earth.

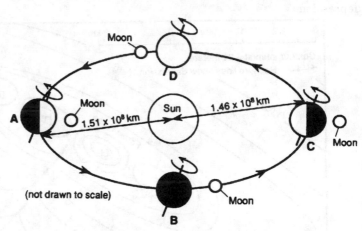

61 What is most likely represented by the symbol ⟲ near the Earth's axis at each position?
1 the direction of Earth's rotation
2 the path of the Sun through the sky
3 the changing tilt of Earth's axis
4 convection currents in the atmosphere

62 The Earth's orbit around the Sun is best described as
1 a perfect circle 3 a very eccentric ellipse
2 an oblate spheroid 4 a slightly eccentric ellipse

63 Which position of the Earth represents the beginning of the winter season for New York State?
(1) A (2) B (3) C (4) D

64 Which position of the Earth shows the Moon located where its shadow may sometimes reach the Earth?
(1) A (2) B (3) C (4) D

65 As the Earth moves from position B to position C, what change will occur in the gravitational attraction between the Earth and the Sun?
1 It will decrease, only. 3 It will decrease, then increase.
2 It will increase, only. 4 It will remain the same.

Group 3
If you choose this group, be sure to answer questions 66-70.

Base your answers to questions 66 through 70 on the *Earth Science Reference Tables*, the weather map of the United States below, and your knowledge of Earth Science.

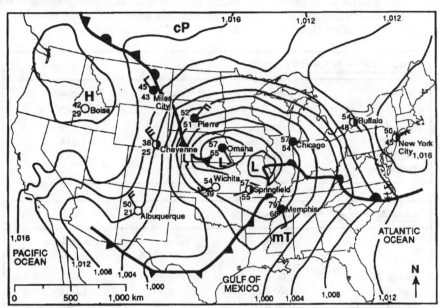

66   Which kind of frontal system is located northwest of Miles City, Montana?
1   cold front   3   stationary front
2   warm front   4   occluded front

67   The air mass over Memphis Tennessee, most likely originated in
1   the North Pacific   3   the central United States
2   central Canada   4   the Gulf of Mexico

68   According to the map, at which city is precipitation most likely occurring?
1   Boise, Idaho   3   Albuquerque, New Mexico
2   Omaha, Nebraska   4   New York City

69   If the low-pressure systems follow the path of most weather systems in the United States, in which direction will they move?
1   northwest   3   northeast
2   southwest   4   southeast

70   The weather front west of Memphis, Tennessee, is moving at a speed of 50 kilometers per hour. What is the most likely weather forecast for Memphis for the next 12 hours?
1   showers followed by clearing skies and cooler temperatures
2   showers followed by warm, humid conditions
3   clearing skies followed by warm, dry conditions
4   a continuation of the present weather conditions

Group 4
If you choose this group, be sure to answer questions 71-75.

Base your answers to questions 71 through 75 on the *Earth Science Reference Tables,* the data table below and your knowledge of Earth Science. The table shows the time of sunrise and sunset and the total amount of insolation received on the Earth's surface for four locations, *A, B, C,* and *D,* at the beginning of each season. The locations have the same longitude, but are at different latitudes. Data were collected on clear, sunny days.

Location A — 66° North Latitude

Date	Sunrise	Sunset	Total Insolation (cal/cm^2)
March 21	6:00 a.m.	6:00 p.m.	373
June 21	12:51 a.m.	11:09 p.m.	1,014
September 23	5:48 a.m.	6:12 p.m.	393
December 21	11:06 p.m.	12:54 a.m.	1

Location B — 43° North Latitude

Date	Sunrise	Sunset	Total Insolation (cal/cm^2)
March 21	6:00 a.m.	6:00 p.m.	674
June 21	4:24 a.m.	7:36 p.m.	1,023
September 23	5:54 a.m.	6:06 p.m.	682
December 21	7:33 a.m.	4:27 p.m.	284

Location C — 0° Latitude

Date	Sunrise	Sunset	Total Insolation (cal/cm^2)
March 21	6:00 a.m.	6:00 p.m.	923
June 21	6:00 a.m.	6:00 p.m.	814
September 23	6:00 a.m.	6:00 p.m.	909
December 21	6:00 a.m.	6:00 p.m.	869

Location D — 23° South Latitude

Date	Sunrise	Sunset	Total Insolation (cal/cm^2)
March 21	5:57 a.m.	6:03 p.m.	851
June 21	6:42 a.m.	5:18 p.m.	545
September 23	6:00 a.m.	6:00 p.m.	827
December 21	5:15 a.m.	6:45 p.m.	1,044

71 Which location received the greatest total insolation on June 21?
    (1) *A*   (3) *C*
    (2) *B*   (4) *D*

72 At which location would an observer see the Sun in the northern sky at
    noon on March 21?
    (1) *A*   (3) *C*
    (2) *B*   (4) *D*

73 A comparison of the times sunrise and sunset in New York State on
    December 21 and June 21 shows that, in December, the Sun
    1 rises later and sets earlier   3 rises and sets earlier
    2 rises earlier and sets later   4 rises and sets later

74 Which statement best explains why surface temperatures are higher at 43°
    N than at 66° N on June 21?
    1 At 66° N, there is complete darkness.
    2 At 66° N, winter is beginning.
    3 At 43° N, summer is ending.
    4 At 43° N, the angle of insolation is greater.

75 Why are the times of sunrise and sunset for March 21 and September 23
    nearly the same at each location?
    1 The Sun's altitude at noon is the same everywhere on Earth on these
       days.
    2 The Earth is at its closest and farthest points from the Sun on these
       days.
    3 The Sun's insolation reaches its maximum intensity on these days.
    4 The Sun is directly above the Equator on these days.

Group 5
If you choose this group, be sure to answer questions 76-80.

Base your answers to questions 76 through 80 on the *Earth Science Reference
Tables*, the diagrams below, and your knowledge of Earth Science. The diagrams
represent cross sections of four samples of loosely packed uniformly sorted soil
particles. The diameter of the particles is given below each diagram. All soil
samples consist of solid spherical particles.

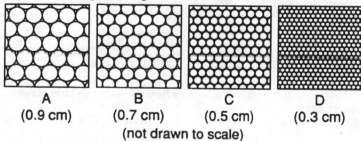

A (0.9 cm)   B (0.7 cm)   C (0.5 cm)   D (0.3 cm)
(not drawn to scale)

76 Particles of the sizes shown are classified as
    1 cobbles   3 sand
    2 pebbles   4 silt

77   Which graph best represents the capillarity of these soil samples?

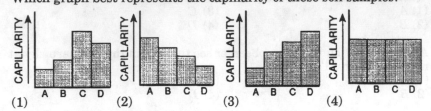

78   Water can infiltrate these soils if they are
1   saturated and impermeable
2   saturated and permeable
3   unsaturated and impermeable
4   unsaturated and permeable

79   Which sample has the greatest permeability?
(1) *A*   (2) *B*   (3) *C*   (4) *D*

Note that question 80 has only three choices.

80   some particles from sample *D* are mixed with particles from sample *A*. Compared to the original porosity of sample *A*, the porosity of the resulting mixture will be
1   less   2   greater   3   the same

Group 6
If you choose this group, be sure to answer questions 81-85.

Base your answers to questions 81 through 85 on the *Earth Science Reference Tables*, the diagram below, and your knowledge of Earth Science. The diagram represents a profile of a stream. Points *A* through *E* are locations along the stream.

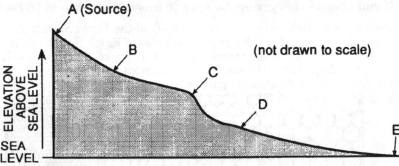

81   The primary force responsible for the flow of water in this stream is
1   solar energy   3   wind
2   magnetic fields   4   gravity

82   Between which two points is potential energy changing to kinetic energy most rapidly?
(1) *A* and *B*   (3) *C* and *D*
(2) *B* and *C*   (4) *D* and *E*

83 The largest particles of sediment transported by the stream at location *C* are sand particles. What is the approximate velocity of the stream at location *C*?
(1) 50 cm/sec   (3) 300 cm/sec
(2) 200 cm/sec   (4) 600 cm/sec

84 In what way would a sediment particle most likely change while it is being transported by the stream?
1 It will become more dense.   3 Its size will decrease.
2 It will become more angular.   4 Its hardness will increase.

85 At which location would the amount of deposition be greatest?
(1) *A*   (2) *B*   (3) *E*   (4) *D*

Group 7
If you choose this group, be sure to answer question 86-90.

Base your answers to questions 86 through 90 on the *Earth Science Reference Tables*, diagrams at the right, and your knowledge of Earth Science.

BASALT

GNEISS
(METAMORPHIC)

CONGLOMERATE   GRANITE   SANDSTONE

86 Which sample is composed of sediments 0.006 centimeter to 0.2 centimeter in size that were compacted and cemented together?
1 conglomerate   3 gneiss
2 sandstone   4 granite

87 If granite were subjected to intense heat and pressure, it would most likely change to
1 conglomerate   3 gneiss
2 sandstone   4 basalt

88 The basalt was most likely formed by
1 heat and pressure   3 compaction and cementation
2 melting and solidification   4 erosion and deposition

89 Which sample would most likely contain fossils?
1 gneiss   3 sandstone
2 granite   4 basalt

90 Which sample is igneous and has a coarse texture?
1 sandstone   3 basalt
2 conglomerate   4 granite

Group 8
If you choose this group, be sure to answer questions 91-95.

Base your answers to questions 91 through 95 on the *Earth Science Reference Tables*, the diagram below, and your knowledge of Earth Science. The diagram shows a cross section of bedrock where the Niagara River flows over Niagara Falls.

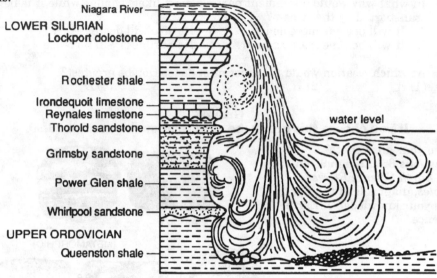

91   The Niagara River begins at
     1   the Genesee River   3   Lake Erie
     2   Niagara Falls   4   the St. Lawrence River

92   Which is the youngest rock unit?
     1   Lockport dolostone   3   Whirlpool sandstone
     2   Queenston shale   4   Rochester shale

93   Which rock layers appear to have weathered and eroded most?
     1   Irondequoit limestone and Whirlpool sandstone
     2   Power Glen shale and Queenston shale
     3   Lockport dolostone and Reynales limestone
     4   Thorold sandstone and Rochester shale

94   A sedimentary layer resembling the Rochester shale is located in another section of New York State. The best way to correlate these two rock units would be to compare the
     1   thickness of the layers
     2   index fossils contained in the layers
     3   minerals cementing the sediments
     4   color of the layers

95   Which rock unit was most likely formed from chemical precipitates?
     1   Lockport dolostone   3   Rochester shale
     2   Whirlpool sandstone   4   Thorold sandstone

Group 9
If you choose this group be sure to answer questions 96-100.

Base your answers to questions 96 through 100 on the *Earth Science Reference Tables*, the map and information below, and your knowledge of Earth Science.

    The map shows the location of major islands and coral reefs in the Hawaiian Island chain. Their ages are given in millions of years.

    The islands of the Hawaiian chain formed from the same source of molten rock, called a hot plume. The movement of the Pacific Plate over the Hawaiian hot plume created a trail of extinct volcanoes that make up the Hawaiian Islands. The island of Hawaii (lower right) is the most recent island formed. Kilauea is an active volcano located over the plume on the island of Hawaii.

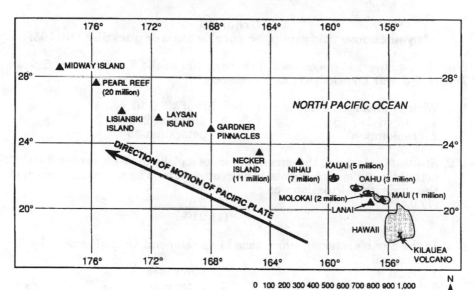

96   Approximately how far has the Pacific Plate moved since Necker Island was located over the hot plume at *X*?
   (1) 300 km   (3) 1,900 km
   (2) 1,100 km   (4) 2,600 km

97   What is the location of Lisianski Island?
   (1) 26° N 174° E   (3) 26° S 174° E
   (2) 26° N 174° W   (4) 26° S 174° W

98   Approximately how long will it take the noon Sun to appear to move from Kauai to Pearl Reef?
   (1) 1 hour   (3) 15 minutes
   (2) 2 hours   (4) 45 minutes

99 What kinds of animals were common in New York State 20 million years ago, when Pearl Reef was forming?
1 primitive humans   3 dinosaurs and ammonites
2 early reptiles   4 grazing mammals

100 Which graph shows the general relationship between the age of individual islands in the Hawaiian chain and their distance from the hot plume?

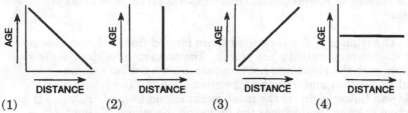

(1)   (2)   (3)   (4)

Group 10
If you choose this group, be sure to answer question 101-105.

Base your answers to questions 101 through 105 on the *Earth Science Reference Tables* and your knowledge of Earth Science.

101 Which radioactive element has a half-life of 4.5 x 10 years?
1 carbon-14   3 uranium-238
2 rubidium-87   4 potassium-40

102 Students calculated the circumference of a globe to be 60. centimeters. The actual circumference of globe is 63 centimeters. The percent deviation of the students' calculation was
(1) 0.48%   (3) 5.0%
(2) 4.8%   (4) 21%

103 In the Earth's interior, which zone has a temperature higher than its melting point?
1 crust   3 inner core
2 stiffer mantle   4 outer core

104 What is the dewpoint temperature when the dry-bulb temperature is 22°C and the wet-bulb temperature is 15°C?
(1) 7°C   (3) 12°C
(2) 10°C   (4) 14°C

105 A pebble has a mass of 35 grams and a volume of 14 cubic centimeters. What is its density?
(1) 0.4 g/cm   (3) 490 g/cm
(2) 2.5 g/cm   (4) 4.0 g/cm

Examination
Earth Science

Pupil _____

School _____

Teacher _____

| Raw Score Total on Part I .... |
| Raw Score Total on Part II .... |
| Raw Score Total on Lab Exam .... |
| Total (Official Regents Examination Mark) _____ |

Part I (55 credits)

1	1 2 3 4	21	1 2 3 4	41	1 2 3 4
2	1 2 3 4	22	1 2 3 4	42	1 2 3 4
3	1 2 3 4	23	1 2 3 4	43	1 2 3 4
4	1 2 3 4	24	1 2 3 4	44	1 2 3 4
5	1 2 3 4	25	1 2 3 4	45	1 2 3 4
6	1 2 3 4	26	1 2 3 4	46	1 2 3 4
7	1 2 3 4	27	1 2 3 4	47	1 2 3 4
8	1 2 3 4	28	1 2 3 4	48	1 2 3 4
9	1 2 3 4	29	1 2 3 4	49	1 2 3 4
10	1 2 3 4	30	1 2 3 4	50	1 2 3 4
11	1 2 3 4	31	1 2 3 4	51	1 2 3 4
12	1 2 3 4	32	1 2 3 4	52	1 2 3 4
13	1 2 3 4	33	1 2 3 4	53	1 2 3 4
14	1 2 3 4	34	1 2 3 4	54	1 2 3 4
15	1 2 3 4	35	1 2 3 4	55	1 2 3 4
16	1 2 3 4	36	1 2 3 4		
17	1 2 3 4	37	1 2 3 4		
18	1 2 3 4	38	1 2 3 4		
19	1 2 3 4	39	1 2 3 4		
20	1 2 3 4	40	1 2 3 4		

Part II (35 credits)
Answer the questions in only seven of the ten groups in this part.

Group 1	
56	1 2 3 4
57	1 2 3 4
58	1 2 3 4
59	1 2 3 4
60	1 2 3 4

Group 2	
61	1 2 3 4
62	1 2 3 4
63	1 2 3 4
64	1 2 3 4
65	1 2 3 4

Group 3	
66	1 2 3 4
67	1 2 3 4
68	1 2 3 4
69	1 2 3 4
70	1 2 3 4

Group 4	
71	1 2 3 4
72	1 2 3 4
73	1 2 3 4
74	1 2 3 4
75	1 2 3 4

Group 5	
76	1 2 3 4
77	1 2 3 4
78	1 2 3 4
79	1 2 3 4
80	1 2 3 4

Group 6	
81	1 2 3 4
82	1 2 3 4
83	1 2 3 4
84	1 2 3 4
85	1 2 3 4

Group 7	
86	1 2 3 4
87	1 2 3 4
88	1 2 3 4
89	1 2 3 4
90	1 2 3 4

Group 8	
91	1 2 3 4
92	1 2 3 4
93	1 2 3 4
94	1 2 3 4
95	1 2 3 4

Group 9	
96	1 2 3 4
97	1 2 3 4
98	1 2 3 4
99	1 2 3 4
100	1 2 3 4

Group 10	
101	1 2 3 4
102	1 2 3 4
103	1 2 3 4
104	1 2 3 4
105	1 2 3 4